NMR: The Toolkit

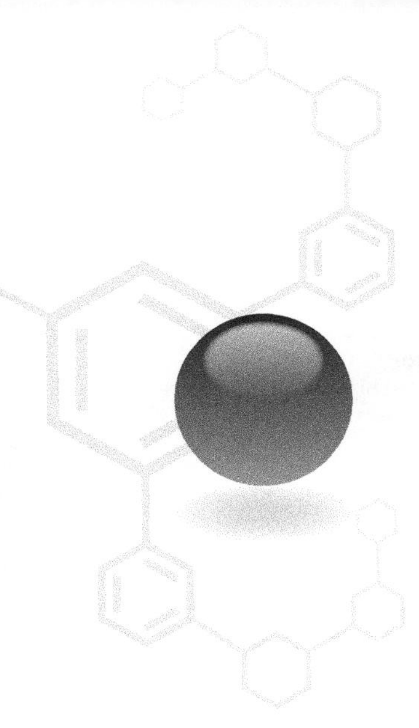

NMR: The Toolkit

How Pulse Sequences Work

SECOND EDITION

P. J. Hore

J. A. Jones

S. Wimperis

Great Clarendon Street, Oxford, OX2 6DP,
United Kingdom

Oxford University Press is a department of the University of Oxford.
It furthers the University's objective of excellence in research, scholarship,
and education by publishing worldwide. Oxford is a registered trade mark of
Oxford University Press in the UK and in certain other countries

First edition 2000

Published in the United States of America by Oxford University Press
198 Madison Avenue, New York, NY 10016, United States of America

British Library Cataloguing in Publication Data
Data available

Library of Congress Control Number: 2015931342

ISBN 978-0-19-870342-6

Printed in Great Britain by Ashford Colour Press Ltd, Gosport, Hampshire

Preface to the 1st edition

The words 'bewildering' and 'baffling' are frequently found in reviews of books on nuclear magnetic resonance (NMR). They usually refer not to the texts themselves but to the vast, perplexing, and ever-expanding array of techniques, applications, pulse sequences, jargon, and, of course, acronyms that characterize modern NMR spectroscopy. To understand and use NMR these days, it is no longer sufficient to be adept at interpreting chemical shifts, spin–spin couplings, and multiplet patterns. One must be comfortable in two, three, or four frequency dimensions, relaxed about experiments on heteronuclear spin systems, excited by multiple quantum coherence, and in tune with sequences of radiofrequency pulses that resemble the sheet music of a Beethoven sonata. This book attempts to explain how some of these experiments work.

Oxford and Exeter P. J. H.
March 2000 J. A. J.
S. W.

Preface to the 2nd edition

In writing this second edition of *The Toolkit* we have resisted the temptation greatly to expand its breadth and depth or to change its ethos. Rather, we have corrected as many of the errors and inaccuracies as possible and have reorganized the bits of the first edition that now seem less than satisfactory. Please let us know if you spot any mistakes or things that are badly explained.

The changes to Part A are relatively minor. Chapter 1 now has a short reminder about spin energy-levels and their relation to spectra, Chapter 2 contains a brief discussion of hypercomplex Fourier transformation and a new section on heteronuclear decoupling, and the product operator treatment of spin echoes in Chapter 3 has been reordered. The section on three-dimensional NMR in Chapter 5 has been expanded and Chapter 6 has a discussion of nested phase cycles.

Part B has seen slightly more modifications, as the easy availability of mathematical software has made it practical to tackle some slightly more complex problems. In Chapter 8, there is a new section on off-resonance pulses, while Chapter 9 has a new section on the use of propagators and the concept of the average Hamiltonian to elucidate spin echoes. The treatment of equivalent spins, and thus of TOCSY, has been moved from Chapter 9 to Chapter 10 so as to follow rather than precede the discussion of strong coupling. The calculation of the free induction decay of a strongly coupled two-spin system in Chapter 10 has been reorganized and a short piece of *Mathematica* code has been added to illustrate how easy it is to do such calculations with a computer. Finally, the appendices, which were sprinkled over a number of chapters, are now collected at the end of the book and there are a few more of them.

Exercises have been added at the end of each chapter. We hope they will help you verify and deepen your understanding of the material in the book. Worked solutions are available on-line at www.oxfordtextbooks.co.uk/orc/hore2e/. Each chapter also has a short summary section. There is a table of experiments and a list of useful computer packages at the end of the book.

We are indebted to Andy Baldwin and Tim Claridge, and especially to Geoffrey Bodenhausen and Philip Kuchel, for detailed comments on the first edition and wise suggestions for improvements. We also thank the other readers who have sent comments or corrections during the last 14 years.

Oxford and Glasgow P. J. H.
October 2014 J. A. J.
S. W.

Preamble

Not another NMR book? Well, yes and no. There are many excellent NMR texts on the market written for everyone from the neophyte to the connoisseur; we hope this one is a bit different. It is intended as a short, approachable description of how modern NMR experiments work, aimed principally at those who use, or might use, an NMR spectrometer and are curious about why the spectra look the way they do. We say little about the practical side of NMR, nor do we discuss applications, both of which are well documented elsewhere. What we hope to do is to provide, in an accessible and relatively informal way, the conceptual and theoretical tools needed to understand the inner workings of some of the multi-pulse, multi-nuclear, multi-dimensional techniques that chemists and biochemists use to probe the structures and dynamics of molecules in liquids. There is no attempt at a comprehensive coverage.

In a sense this is two books in one, going over similar ground in different ways. In principle, one could read either independently of the other, although it would probably be simpler to begin with Chapter 1. Part A (Chapters 1–6) starts with the *vector model*, a pictorial description of simple NMR experiments, and proceeds to the more powerful *product operator formalism* with which one can appreciate the mechanics of many complex pulse sequences and predict the spectra they produce. Having discussed some quite sophisticated techniques towards the end of Part A, we go back to basics in Part B (Chapters 7–10) and show how straightforward *quantum mechanics* can be used to understand NMR at a more fundamental level. Amongst other things, Part B attempts to show what product operators really are, and to provide justifications for some of the ideas and results we ask the reader to take on trust in Part A. It also shows how to handle cases that are beyond the scope of product operators.

We assume the reader is broadly familiar with the fundamental interactions that control the appearance of simple liquid-state NMR spectra, namely chemical shifts and spin–spin couplings. If you need to remind yourself of these topics, we recommend Chapters 1–3 of the Oxford Chemistry Primer *Nuclear Magnetic Resonance* by P. J. Hore, which is also available now in a second edition. Chapter 6 of that book overlaps with Chapter 1 of *The Toolkit* and might be useful if you find our presentation of the vector model too compact or would like a different introduction to two-dimensional NMR. The two books now use a consistent convention for the signs of Larmor frequencies and the rotations produced by radiofrequency pulses.

Although the treatment is of necessity mathematical, we have tried to keep things as simple as is consistent with a reasonable level of accuracy. The margins carry brief reminders of bits and pieces of algebra, which the more mathematically sophisticated will probably wish to ignore. The appendices contain material that would disrupt the flow of the text.

Contents

Part A

Product Operators

The vector model

1.1 Introduction

The so-called *vector model* of NMR spectroscopy is an essential weapon in the armoury of every practising NMR spectroscopist because it provides the sort of simple, intuitive, non-mathematical picture that every human brain eventually requires. Although it is an excellent way of understanding NMR experiments on *isolated* spin-$\frac{1}{2}$ nuclei, the vector model has only limited applicability for interacting or 'coupled' spins, and cannot be used at all to appreciate the inner workings of many important NMR experiments. Nevertheless, it is fundamental to NMR and forms the basis for a much more versatile and powerful formalism, the *product operator* description, which will be introduced in Chapters 3–5. We assume the reader has some acquaintance with elementary NMR and the vector model, which we now review.

A more detailed treatment of these elementary topics can be found in Hore *Nuclear Magnetic Resonance* (2015).

A spin-$\frac{1}{2}$ nucleus is a nucleus with spin quantum number $I=\frac{1}{2}$, as discussed in Section 1.2.

1.2 Bulk magnetization

Elementary quantum mechanics reveals that an atomic nucleus with nuclear *spin quantum number* I in a magnetic field $\boldsymbol{B}_0$ has $2I+1$ non-degenerate energy levels, which can be distinguished by their *magnetic quantum number* m, which runs from I down to $-I$ in steps of 1. Neglecting the *chemical shift*, which will be discussed in Section 1.5, and J-coupling, which occurs in systems of more than one spin, the energies of these levels are given by

$$E(m) = -\hbar\gamma B_0 m = \hbar\omega_0 m \tag{1.1}$$

where $\omega_0 = -\gamma B_0$, called the *Larmor frequency*, depends on the magnetic field strength and on γ, the *magnetogyric ratio*, which is an intrinsic property of the nuclide. Note that ω_0 is an *angular* frequency, measured in radians per second, and must be divided by 2π to obtain a conventional frequency, $\nu_0 = \omega_0/2\pi$, measured in hertz (Hz). Transitions can occur between these energy levels according to the selection rule $\Delta m = \pm 1$. The energies of these transitions are equal to the gaps between adjacent levels, all of which are identical: $\Delta E = |E(m+1) - E(m)| = \hbar|\omega_0|$.

For example, a ^{1}H nucleus (or proton) has spin quantum number $I=\frac{1}{2}$ and is found to occupy one of two distinct energy levels, often labelled α ($m=+\frac{1}{2}$) and β ($m=-\frac{1}{2}$), with a separation of about 400 MHz when B_0 = 9.4 T. The classical

The spin quantum number I can be integral or half-integral, i.e. it can take values $I = 0, \frac{1}{2}, 1, \frac{3}{2}, \ldots$ depending on the nuclide.

We use ***bold italic type*** here to denote vector quantities, which have both an amplitude and direction. Often the same symbol in *italic type* is used for the length of the vector. Thus, B_0 denotes the strength of the magnetic field $\boldsymbol{B}_0$.

The magnetogyric ratio is sometimes called the gyromagnetic ratio. Note that for nuclei with a positive value of γ (the more common case) the corresponding Larmor frequency is *negative*. In many cases this fact can be safely neglected, and we will generally choose to do so. For a more careful treatment see Levitt *The Signs of Frequencies and Phases in NMR* (1997).

From here on we will only consider nuclei with $I=\frac{1}{2}$, colloquially known as spin-$\frac{1}{2}$ nuclei. Important examples include ^{1}H, ^{13}C, ^{15}N, ^{19}F, and ^{31}P.

The tesla (T) is the SI unit for magnetic field strength (magnetic induction), and 1 tesla is equivalent to 10^4 gauss (G).

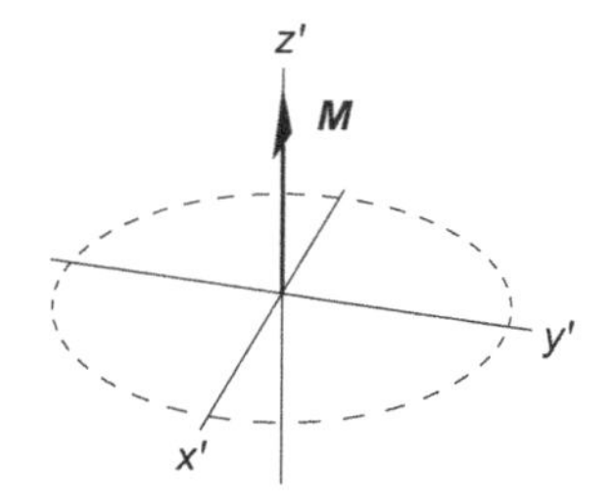

Fig. 1.1 The bulk magnetization $\boldsymbol{M}$ in the laboratory frame at thermal equilibrium.

picture of this situation is that the nuclear magnetic moment is precessing in the magnetic field $\boldsymbol{B}_0$ at the Larmor frequency, and has its precession axis aligned either parallel (low energy) or antiparallel (high energy) to $\boldsymbol{B}_0$. In a macroscopic sample at thermal equilibrium there will be a very slight excess of spins in the lower energy level, which leads to a net or *bulk magnetization* $\boldsymbol{M}$ of the sample; $\boldsymbol{M}$ is stationary and aligned parallel to $\boldsymbol{B}_0$ (Fig. 1.1). The direction of the magnetic field $\boldsymbol{B}_0$ defines the z'-axis of an x', y', z' coordinate system, which is known as the *laboratory frame*; the x' and y' axes are (arbitrarily) fixed in space. (The reason for putting 'primes' on x', y', and z' will soon become clear.)

The size of this bulk magnetization depends on the intrinsic magnetization of the nuclear spin and the distribution of the spins between the $m = \pm\frac{1}{2}$ energy levels, which can be calculated using the Boltzmann distribution. As the energy gap is small compared with $k_B T$ at room temperature, it is sensible to use a high-temperature approximation to estimate the *polarization*

The intrinsic magnetization of the ^{1}H nuclei is given by $M = \frac{1}{2}\gamma\hbar\Delta n_{eq}$ and $k_B \approx 1.38 \times 10^{-23}$ J K^{-1} is the Boltzmann constant.

$$\frac{\Delta n_{eq}}{n_{total}} = \frac{n_{lower} - n_{upper}}{n_{lower} + n_{upper}} \approx \frac{\Delta E}{2k_B T} \qquad (1.2)$$

which gives a value of about 3.2×10^{-5} for ^{1}H at 9.4 T and 298 K.

1.3 The rotating frame

The bulk magnetization can be rotated away from its equilibrium position by the effects of a *pulse*, which is a linearly oscillating magnetic field applied for a short time along (say) the x'-axis of the laboratory frame. As the frequencies of these oscillating fields are typically several hundred megahertz, it is common to refer to them as *radiofrequency* (rf) *fields* and *radiofrequency pulses*. The oscillation frequency ω_{rf} (the *transmitter frequency*) is chosen to lie very close to the Larmor frequency ω_0, also called the *resonance frequency* of the spins. The effect of a nearly resonant pulse is to tilt the magnetization vector $\boldsymbol{M}$ away from the z'-axis. Once this motion starts, however, $\boldsymbol{M}$ will also immediately begin to precess about the z'-axis (i.e. $\boldsymbol{B}_0$) at the Larmor frequency. These superimposed motions are difficult to visualize because of their rapid and complicated time dependence.

The problem can be simplified by thinking of the linearly oscillating radiofrequency field as the sum of two counter-rotating fields with angular frequencies ω_{rf} and $-\omega_{rf}$. Only the component that rotates in the same sense as the Larmor precession (ω_{rf}) is retained; the other ($-\omega_{rf}$) is far off-resonance and has little effect on the spins. If the NMR experiment is now viewed in a *rotating frame*, which rotates about the z'-axis with angular frequency ω_{rf}, the rotating radiofrequency field appears to be static. Thus, in the rotating frame one can view a pulse simply as the temporary application of a static magnetic field $\boldsymbol{B}_1$, orthogonal to $\boldsymbol{B}_0$. The other consequence of moving into the rotating frame is that the static field $\boldsymbol{B}_0$ is replaced by an *offset* field $\Delta\boldsymbol{B}_0$ along the z'-axis (see Section 1.5). In many situations the radiofrequency field $\boldsymbol{B}_1$, when present, is strong enough that $\Delta\boldsymbol{B}_0$ can be ignored. The axes of the rotating frame are simply labelled x, y, and z, with the last corresponding to the z'-axis of the laboratory frame.

The component at $-\omega_{rf}$ gives rise to a small (usually negligible) change in the resonance frequency called the *Bloch–Siegert shift*.

As the rotating frame rotates along with the spin, the spin's apparent precession frequency is reduced, and so from Eqn 1.1 the apparent field must also be reduced. This argument is more formally justified in Appendix F.

1.4 Nutation

In the rotating frame the effect of a radiofrequency pulse is simple: once the pulse is turned on, the bulk magnetization 'sees' an apparently static magnetic field $\boldsymbol{B}_1$ and precesses (or *nutates*) about it until it is turned off (Fig. 1.2).

Equivalent processes also occur in other types of coherent spectroscopy, and spin nutation in NMR is an example of a general quantum mechanical process called *Rabi flopping*, as discussed in Jones and Jaksch *Quantum Information, Computation and Communication* (2012).

Here, and in most of the rest of this book, we assume that the radiofrequency pulse is perfectly on-resonance with the transition. The treatment of off-resonance pulses is briefly addressed in Chapter 8.

Note that, contrary to what is found in some elementary texts, the axes and the sense of nutation have been defined such that a pulse about the $+x$-axis takes $\boldsymbol{M}$ initially towards the $-y$-axis. The rate or angular frequency of precession (the *nutation frequency*) is given by $\omega_1 = -\gamma B_1$. The angle β through which $\boldsymbol{M}$ nutates is given by $\beta = \omega_1 t_p$, where t_p is the duration of the pulse or *pulse length*. This time can be chosen to make the *flip angle* β equal to $\pi/2$ (a 90° pulse) if maximum excitation of transverse magnetization is required, or π (a 180° pulse) to invert the equilibrium magnetization (Fig. 1.3).

It does not really matter which sign convention is adopted, provided everything is done consistently. The different conventions account for the minor discrepancies between equations and figures here and in some other texts.

As noted above, for a nucleus with a positive value of γ the nutation frequency ω_1 must in fact be negative. See Levitt (1997) for a detailed discussion of this problem, which can be handled in several different ways. Here we simply ignore it.

Most modern spectrometers can apply pulses about any axis in the xy plane of the rotating frame. In many experiments, however, only pulses about the $+x$, $+y$, $-x$, and $-y$ axes are used. This shifting of the direction of the pulse axis in the rotating frame is achieved by altering the *phase* of the radiofrequency field in the laboratory frame and not, of course, by any physical movement of the radiofrequency coil. It should also be remembered that phase is not an absolute term: when we talk of the phase of a pulse we actually mean the phase of the radiofrequency field relative to a reference, e.g. the phase of the reference frequency of the detector (see Chapter 2).

Such pulses are conveniently labelled by their nutation angle and axis. For example, a 90°_x pulse causes a nutation of 90° around the x-axis of the rotating frame.

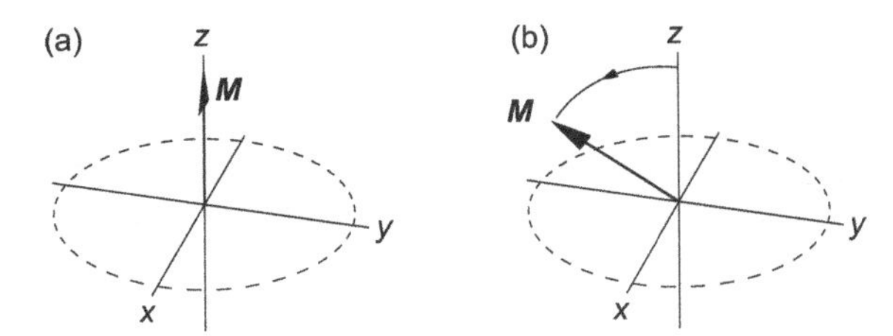

Fig. 1.2 The effect of a radiofrequency pulse. (a) Before the pulse, $\boldsymbol{M}$ is aligned along the z-axis of the rotating frame. (b) During the pulse, $\boldsymbol{M}$ precesses about the field $\boldsymbol{B}_1$, which is aligned along the x-axis.

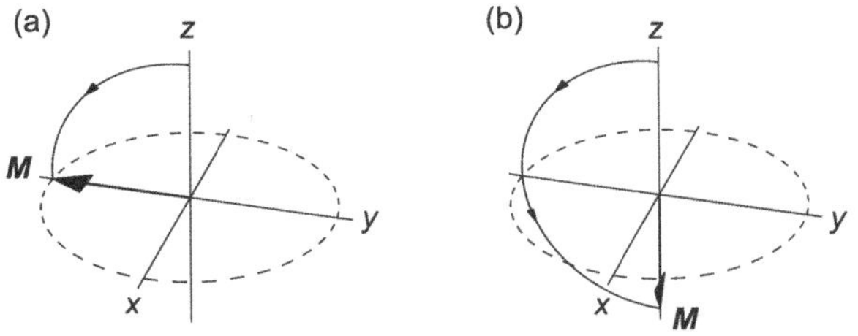

Fig. 1.3 The bulk magnetization, $\boldsymbol{M}$, (a) after a 90° pulse and (b) after a 180° pulse, both about the x-axis in the rotating frame.

1.5 Free precession

If the NMR spectrum consists of a single peak, e.g. the ^1H spectrum of water, then the transmitter frequency can be set precisely equal to the Larmor frequency, $\omega_{rf} = \omega_0$. A 90°_x pulse rotates the bulk magnetization $\boldsymbol{M}$ to the $-y$-axis, but it does *not* then start to precess about the z-axis. This is because the transformation from the laboratory frame to the rotating frame has, in this case, effectively removed the static magnetic field $\boldsymbol{B}_0$. The offset field ΔB_0 along the z-axis in the rotating frame is

The origin of the offset field in the rotating frame is covered in Appendix F.

$$\Delta B_0 = B_0 + \omega_{rf}/\gamma = -(\omega_0 - \omega_{rf})/\gamma \tag{1.3}$$

which vanishes if $\omega_{rf} = \omega_0$, i.e. if the pulse is exactly 'on-resonance'.

The actual Larmor frequency is not, however, given precisely by the formula in Section 1.2, but rather by

$$\omega_0 = -\gamma B_0 (1 - \sigma) \tag{1.4}$$

where σ, the *shielding* or *screening constant*, depends on the local chemical environment of the nucleus. This leads to different resonance frequencies for nuclei in different environments. As these variations in frequency are proportional to B_0 it is convenient to describe these *chemical shifts* by fractional shifts from some reference frequency, and they are usually measured in *parts per million*, or ppm. The typical range of chemical shifts is quite small: around 10 ppm for ^1H and about 200 ppm for ^{13}C.

Because of these narrow frequency ranges, NMR signals from different nuclides almost never overlap.

As a result of chemical shifts, the *resonance offset*, $\Omega = \omega_0 - \omega_{rf}$, is in general non-zero, so that there will be a residual field $\Delta B_0 = -\Omega/\gamma$ in the rotating frame and $\boldsymbol{M}$ will start to precess about it as soon as the pulse is switched off. The sense and frequency of precession are governed by the sign and magnitude of Ω (Fig. 1.4). The angle through which $\boldsymbol{M}$ precesses in the rotating frame in a time t is Ωt radians.

Note that the sign of Ω need not, of course, be the same as the sign of ω_0. In the schematic spectra in Fig. 1.4 the frequency axis has been drawn such that frequencies increase from left to right.

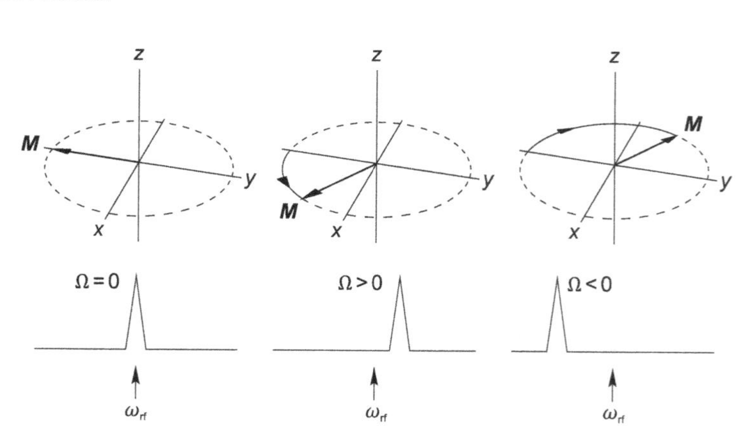

Fig. 1.4 The sense and frequency of precession of $\boldsymbol{M}$ in the rotating frame depend on the magnitude and sign of the resonance offset Ω: (a) $\Omega = 0$ (on-resonance); (b) $\Omega > 0$; (c) $\Omega < 0$. Schematic NMR spectra are shown in the lower part of the figure.

Fig. 1.5 Schematic NMR spectrum of a weakly coupled homonuclear system of two spin-$\frac{1}{2}$ nuclei, I and S. The offset frequencies in the rotating frame are given by $\Omega_I = \omega_{0I} - \omega_{rf}$ and similarly for Ω_S.

In systems with two or more nuclear spins it is also necessary to consider the effects of spin-spin couplings, often called *scalar* couplings or *J*-couplings, which cause the energy of an NMR transition to depend on the spin-states of neighbouring nuclei. For a molecule with two spin-$\frac{1}{2}$ nuclei, conventionally called I and S, the energy levels depend not only on the two Larmor frequencies, ω_{0I} and ω_{0S}, but also on the coupling constant.

The use of the two letters I and S is traditional. Here $\omega_{0I} = -\gamma_I B_0(1 - \sigma_I)$ and similarly for ω_{0S}. Note that coupling constants are traditionally measured in Hz, so it is necessary to multiply by 2π to obtain angular frequencies.

The situation is simplest in the *weak coupling* limit, which applies when the coupling is small in comparison with the difference between the Larmor frequencies of the two nuclei, i.e. $|2\pi J_{IS}| \ll |\omega_{0I} - \omega_{0S}|$. This condition is always true when I and S are different nuclides (*heteronuclear*, e.g. ^{1}H and ^{13}C), but may not be true if they are the same nuclide (*homonuclear*, e.g. two protons). In the weak coupling limit the energy levels are given by

In Part A we assume that all couplings are weak; strong couplings will be discussed in Part B.

$$E(m_I, m_S) = \hbar\omega_{0I} m_I + \hbar\omega_{0S} m_S + \hbar 2\pi J_{IS} m_I m_S. \tag{1.5}$$

The selection rule is that nuclei undergo transitions independently of their coupling partners, such that $\Delta m_I = \pm 1$ or $\Delta m_S = \pm 1$. This results in a spectrum (Fig. 1.5) containing four lines, two lines (a doublet) centred at ω_{0I}, and two more (another doublet) at ω_{0S}, with each doublet split by $2\pi J_{IS}$.

1.6 T_1 and T_2 relaxation

The magnetization $\boldsymbol{M}$ produced, for example, by a 90° pulse does not precess in the *xy* plane indefinitely. The populations of the energy levels, which are *equal* immediately after a 90° pulse, typically return to thermal equilibrium—the Boltzmann distribution—in a few seconds (although in some samples it can take microseconds, milliseconds, minutes, or even hours). This process is known as *spin-lattice* or *longitudinal relaxation*. Very often it is approximately exponential and can be characterized by a single time constant, T_1. Longitudinal relaxation is usually measured with the *inversion recovery* method: a 180° pulse aligns $\boldsymbol{M}$ along the $-z$-axis and then, after a variable time, τ, a 90° pulse monitors the amount of magnetization present. By repeating the experiment for a range of values of τ, the T_1 time constant can be determined.

Spin relaxation is in general far more complex than this very brief discussion suggests. The underlying ideas are discussed in a little more detail in Hore (2015), but a full understanding requires a detailed quantum mechanical treatment of the problem; see, for example, Levitt *Spin Dynamics* (2008).

Note that in many experiments the pulses are so short that one can ignore relaxation while the radiofrequency field is present.

Magnetization in the *xy*-plane (transverse magnetization) created by a pulse also decays with time, but in this case back to its equilibrium value of zero. This process, called *spin–spin* or *transverse relaxation*, is often described as a loss of *phase coherence* between the individual spins. Usually it is approximately exponential and can be characterized by a single time constant, T_2.

1.7 Spin echoes

Although spin echoes are best known in NMR, related phenomena are seen in other types of coherent spectroscopy. See Jones and Jaksch (2012) for a brief discussion of echoes as a general quantum mechanical phenomenon.

The vector model is ideal for explaining the classic NMR phenomenon of the *spin echo*. Imagine an NMR spectrum consisting of a single peak, e.g. the ^{1}H spectrum of water. A 90°_x pulse aligns $\boldsymbol{M}$ along the $-y$-axis (Fig. 1.6). The magnetization then precesses freely for a time τ before a 180°_y pulse flips $\boldsymbol{M}$ to its mirror image position on the other side of the y-axis. Free precession for an identical period τ then causes $\boldsymbol{M}$ once again to be aligned along the $-y$-axis, irrespective of the resonance offset Ω. The 180° pulse is said to *refocus* the chemical shift or resonance offset of the water, and is often referred to as a *refocusing* pulse. If there are many resonances in the spectrum, all their chemical shifts will be refocused along the same axis by the spin echo.

But what happens when the NMR sample contains a pair of interacting (J-coupled) spins? The answer depends on whether the two nuclei are of the same species or not.

Let us start with the easier heteronuclear case and focus on what happens to the magnetization of a ^{1}H nucleus coupled to a ^{13}C. The normal ^{1}H NMR spectrum is a doublet, centred at the ^{1}H resonance offset Ω with splitting $2\pi J$. The two components of the doublet correspond to the α and β spin orientations of the ^{13}C coupling partner. There are therefore two ^{1}H magnetization vectors, with precession frequencies $\Omega \pm \pi J$. A spin echo experiment, with radiofrequency pulses at the ^{1}H NMR frequency, leads to complete refocusing of the heteronuclear J-coupling as well as the offset, as shown in Fig. 1.7.

As usual the spin–spin coupling constant J is quoted in hertz rather than radians per second, hence the 2π.

In contrast, homonuclear J-couplings are *not* refocused because the α and β spin states of the coupling partner are exchanged by the 180° pulse at the same time as the two magnetization vectors are flipped across the xy plane (Fig. 1.8). The angle between the two vectors continues to grow and at the end of the 2τ period is $(2\pi J)\times(2\tau) = 4\pi J\tau$. The echo amplitude is said to be *modulated* by the J-coupling: the two vectors are returned to the $-y$-axis if $\tau = 1/J$.

If an experiment requires that the heteronuclear coupling *not* be refocused then 180° pulses must be applied to both nuclei simultaneously; for the case of a ^{1}H nucleus coupled to a ^{13}C, an additional 180° pulse must be applied to the ^{13}C spins at the same time as the proton 180° pulse. Thus, it is possible to make heteronuclear couplings behave like homonuclear couplings; the reverse

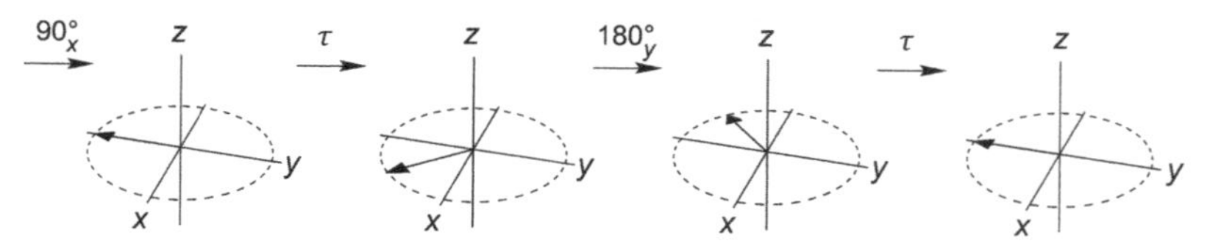

Fig. 1.6 A spin echo for an isolated nucleus (no J-coupling). The chemical shift/resonance offset is refocused after the second period of free precession.

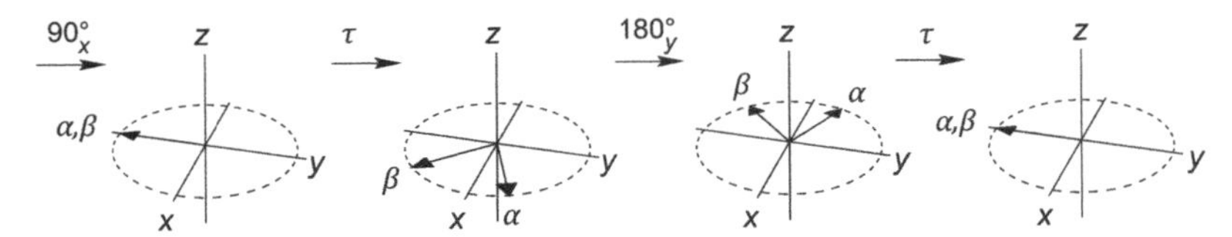

Fig. 1.7 A spin echo for a nucleus with *heteronuclear J*-coupling to a spin-$\frac{1}{2}$ partner. The two magnetization vectors represent the components of the NMR doublet; α and β denote the spin orientations of the coupling partner. Both the chemical shift and the *J*-coupling are refocused at time 2τ.

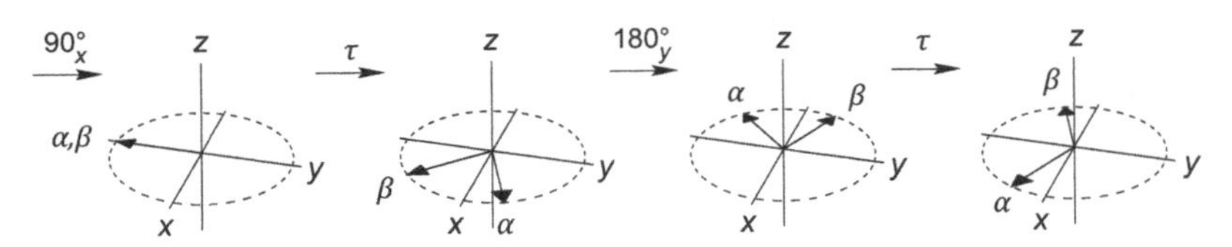

Fig. 1.8 A spin echo for a nucleus with homonuclear *J*-coupling to a spin-$\frac{1}{2}$ partner. Now the *J*-coupling is not refocused because the 180° pulse also inverts the partner spin and interchanges the positions of the α and β magnetization vectors.

process, making homonuclear couplings behave like heteronuclear couplings, is more difficult, but in some cases can be achieved using *frequency-selective pulses.*

We return to spin echo modulation in Chapter 3; as we shall see, it is a crucial element in a large number of NMR experiments.

The use of frequency-selective pulses will be briefly discussed in Chapter 5.

1.8 Summary

- The vector model provides a simple intuitive picture of NMR experiments on isolated spins.
- It can be used to describe the effects of pulses, periods of free precession, and spin relaxation.
- The rotating frame is a crucial simplifying feature of the vector model.
- There are two types of spin relaxation, with characteristic times T_1 and T_2.
- The vector model is less useful for coupled spins, but can be used to understand spin echoes.
- Chemical shifts are refocused in spin echo experiments.
- Homonuclear *J*-couplings lead to echo modulation.
- Heteronuclear *J*-couplings can either be refocused or cause echo modulation depending on how the experiment is performed.

1.9 Exercises

Worked solutions to the exercises are available on the Online Resource Centre at www.oxfordtextbooks.co.uk/orc/hore2e/

1.1. A radiofrequency pulse is applied to a sample of isolated spin-$\frac{1}{2}$ nuclei in thermal equilibrium with spin state populations n_α (lower level) and n_β

(upper level). How will these populations be changed if the pulse flip angle β is: (a) 90°, (b) 180°, and (c) 45°?

1.2. The z-magnetization of a sample of isolated nuclear spins recovers after a 90° pulse according to $M_z = M_0[1 - \exp(-t/T_1)]$. For how many multiples of T_1 is it necessary to wait for $M_z(t)$ to recover to within 1% of its equilibrium value?

1.3. Write an equation for the recovery of $M_z(t)$ after a 180° pulse. At what time does $M_z(t)$ change sign from negative to positive?

1.4. In a $90^\circ_x - \tau - 180^\circ_y - \tau -$ spin echo experiment on a heteronuclear two-spin system (I and S), with pulses applied only to spin I, the magnetization of spin I is refocused along the $-y$-axis. What happens when the phase of the 180° pulse is x instead of y, i.e. $90^\circ_x - \tau - 180^\circ_x - \tau -$?

1.5. In a $90^\circ_x - \tau - 180^\circ_y - \tau -$ spin echo experiment on a homonuclear two-spin system with a J-coupling of 10 Hz, what delay τ is needed to obtain a phase difference of π radians between the two components of the I-spin (or S-spin) doublet at time 2τ?

Fourier transform NMR

2.1 Introduction

This chapter reviews some of the basic elements of Fourier transform NMR spectroscopy, including a brief introduction to two-dimensional NMR. We shall not refer to specific experiments, but will attempt to lay the foundations for the more detailed discussions that follow. As with the previous chapter, the material should be at least vaguely familiar to anyone who uses, or is learning to use, a modern NMR spectrometer.

2.2 Detection of the NMR signal

As described in Chapter 1, a pulse creates magnetization that precesses in the $x'y'$-plane of the laboratory frame. This magnetization is detected by the coil in the NMR probe and constitutes the *free induction decay*, often abbreviated as FID. All the frequencies detected, however, will be of the form $\omega_{rf} + \Omega$, where the transmitter frequency $\nu_{rf} = \omega_{rf}/2\pi$ is typically several hundred megahertz while the resonance offsets $\Omega/2\pi$ are of the order of the chemical shift range of the nucleus being studied, typically a few kilohertz at most. The signal detected in the coil is 'mixed' with the basic spectrometer reference frequency (ω_{rf}) so that only the offset frequencies Ω are passed to the analogue-to-digital converter for digitization. The result of mixing the detected signal with ω_{rf} is precisely equivalent to the spectrometer observing the magnetization in a reference frame rotating at ω_{rf}. The rotating frame is therefore not just a convenient mathematical fiction; it actually coincides neatly with the experimental reality. The spectrometer reference frequency has a defined phase so detection is along a fixed axis of the rotating frame.

ω_{rf}, like ω_0 and Ω, is an *angular* frequency, measured in radians per second.

This mixing process is often performed in two stages, first mixing down from the NMR frequency to some *intermediate frequency*, typically around 20 MHz, and then mixing down again to *audio frequencies*, below 100 kHz. Some modern spectrometers digitize the signal directly at the intermediate frequency. These experimental details do not alter the underlying concepts.

The free induction decay is the sum of many oscillating waves of differing frequencies, amplitudes, and phases. It is detected using two orthogonal (in the rotating frame) detection channels along (say) the x- and y-axes. This is known as *quadrature detection* and it is used by many modern NMR spectrometers. For each resonance in the spectrum, the two signals so acquired are cosine and sine functions of the offset frequency Ω, decaying at a rate $1/T_2$, as indicated in

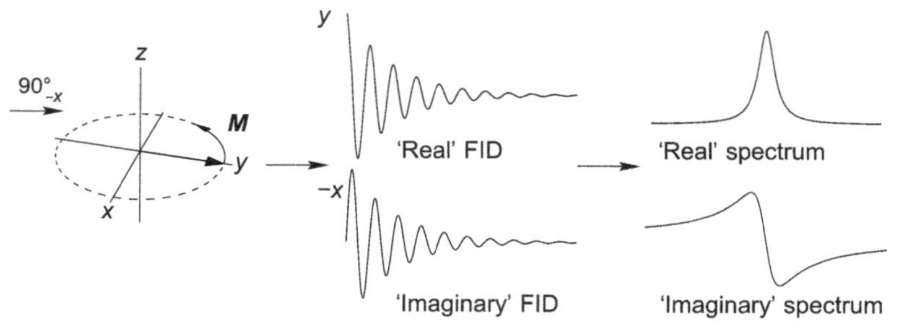

Fig. 2.1 Quadrature detection: simultaneous measurement of two components of the free induction decay, in this case taken as the *y* and –*x* components, using detectors fed with reference signals that differ in phase by 90°.

Fig. 2.1. They can be regarded as the real and imaginary parts of a complex time-domain signal *s*(*t*) of the general form

$$\begin{aligned} s(t) &= [\cos\Omega t + i\sin\Omega t]\exp(-t/T_2) \\ &= \exp(i\Omega t)\exp(-t/T_2) \qquad t \geq 0 \\ s(t) &= 0 \qquad t < 0 \end{aligned} \tag{2.1}$$

N.B. $i = \sqrt{-1}$ and $\exp(iA) = \cos A + i\sin A$

which can be converted into the frequency-domain function or *spectrum* $S(\omega)$ by Fourier transformation:

$$S(\omega) = \int_{-\infty}^{\infty} s(t)\exp(-i\omega t)dt. \tag{2.2}$$

Thus,

$$S(\omega) = A(\Delta\omega) - iD(\Delta\omega) \tag{2.3}$$

with

$$A(\Delta\omega) = \frac{1/T_2}{(1/T_2)^2 + (\Delta\omega)^2} \quad \text{and} \quad D(\Delta\omega) = \frac{\Delta\omega}{(1/T_2)^2 + (\Delta\omega)^2}. \tag{2.4}$$

The frequency parameter $\Delta\omega = \omega - \Omega$ is defined with respect to the centre of the resonance at $\omega = \Omega$. The real part, $A(\Delta\omega)$, of the spectrum is an *absorptive* Lorentzian curve, centred on frequency Ω with a full-width at half height of $1/\pi T_2$ (measured in Hz), while the imaginary part, $D(\Delta\omega)$, is the corresponding *dispersive* Lorentzian (Fig. 2.1). In practice, only the real part of the spectrum is retained: absorptive lines are narrower than their dispersive counterparts and have their maximum amplitude at the frequency of interest.

If only a single detector is used then the spectrometer reference frequency must be placed outside the range of possible frequencies in the FID as their signs are then unambiguous. Quadrature detection is favoured as the transmitter frequency can be placed near the centre of the spectrum, thus making the most efficient use of the available radiofrequency power.

If detection is just carried out along one axis of the rotating frame, it is impossible to determine the sense of precession of the magnetization vectors, i.e. the signs of the frequencies in the NMR spectrum. This can be seen by replacing the $\exp(i\Omega t)$ term in Eqn 2.1 by

$$\cos(\Omega t) = \frac{\exp(i\Omega t) + \exp(-i\Omega t)}{2} \tag{2.5}$$

or

$$\sin(\Omega t) = \frac{\exp(i\Omega t) - \exp(-i\Omega t)}{2i} \tag{2.6}$$

and Fourier transforming. Quadrature detection allows the spectrometer reference frequency, ω_{rf}, to be placed in the centre of the spectrum.

In reality, the NMR spectrum obtained by Fourier transforming the FID rarely turns out as in Fig. 2.1, with the absorption lineshape in the real part of the spectrum and the dispersion lineshape in the imaginary part. The two main reasons for this are (i) it is only by coincidence that the pulse aligns the initial magnetization with the real channel of the detector, and (ii) the pulse has finite length and the spectrometer always inserts a short delay between the end of the pulse and the start of data acquisition; during both these times the various magnetization vectors develop phases that vary linearly with their offset frequency. So, immediately after Fourier transformation, a typical line in the real part of an NMR spectrum has arbitrary phase. Therefore, in order to obtain an absorption lineshape in the real part of the spectrum, one must 'phase the spectrum', which involves taking linear combinations of the real and imaginary parts until the desired result is achieved (Fig. 2.2).

Although the spectrum $S(\omega)$ above was obtained analytically, this is not how the Fourier transform is done in the spectrometer computer. The free induction decay is first converted from analogue to digital form by sampling the oscillating signal at equal time intervals before the spectrum is calculated using an efficient numerical Fourier transform algorithm. The interval between samples Δt is determined by the *Nyquist condition*, $\Delta t = 1/SW$, where SW is the desired spectral width, i.e. the spectrum extends $SW/2$ either side of the reference frequency. This corresponds to performing at least two samples per cycle of every oscillation present in the free induction decay.

For a detailed discussion of numerical data processing in NMR see Hoch and Stern *NMR Data Processing* (1996).

A crucial feature of Fourier transformation is that the process is *linear*, which means that the Fourier transform of the sum of two (or more) signals is just the sum of the Fourier transforms of the separate signals. The free induction decay from a complex molecule will contain many different signals, each of which gives rise to its own line in the NMR spectrum. For most purposes it is sufficient to consider the behaviour of a single signal, and then, at the end, generalize to the case of many signals.

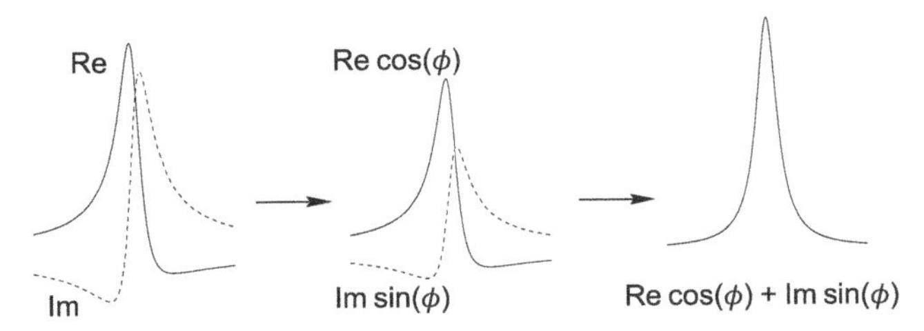

Fig. 2.2 Complex Fourier transformation of a free induction decay with quadrature detection yields a spectrum in need of phase correction. A linear combination of the real (Re) and imaginary (Im) parts of the spectrum gives absorption phase lineshapes in the real part.

2.3 Two-dimensional NMR

According to ancient tradition, the pulse sequence for a generic two-dimensional experiment is broken down into four sequential steps:

Preparation – Evolution – Mixing – Detection.

The meaning of the term *mixing* will become apparent later.

The free induction decay is acquired in the detection period, the time axis of which is normally labelled t_2. In most of the two-dimensional experiments that will concern us, the evolution period consists of uninterrupted free precession for a time, t_1. Hence one often talks of the t_1 *period* and the t_2 *period* of a two-dimensional experiment. The most basic form that the preparation period can take is a single 90°_x pulse to excite transverse magnetization. Similarly, one of the simplest mixing steps is also a 90°_x pulse. This pulse sequence is illustrated in Fig. 2.3.

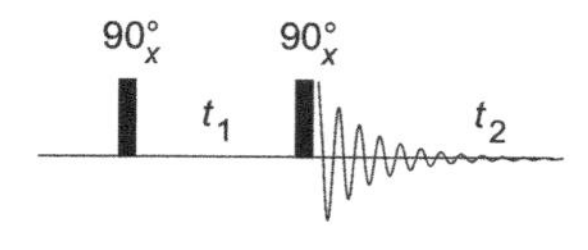

Fig. 2.3 A simple $90^\circ_x - t_1 - 90^\circ_x - t_2$ pulse sequence for two-dimensional NMR.

Once again, imagine an NMR spectrum consisting of a single peak. During t_1 the magnetization vector precesses at a frequency given by the offset Ω until the second 90°_x pulse flips the vector into the *xz*-plane (Fig. 2.4). The component of $\boldsymbol{M}$ that is left along the *z*-axis after the second 90° pulse does not contribute to the signal observed in t_2.

Using a two-dimensional NMR experiment to study a single isolated spin as described here is essentially pointless, but provides a useful introduction to the underlying ideas. The linearity of the Fourier transform ensures that these basic ideas can also be applied in more complex systems with multiple interacting spins.

Although two-dimensional spectroscopy is best known in NMR, where it originated, similar experiments are now used in other types of coherent spectroscopy.

As indicated in Fig. 2.5(a), the amplitude of the observed free induction decay (the *x*-component of $\boldsymbol{M}$ after the pulse) depends on the duration of the evolution period t_1. Therefore a series of experiments performed at increasing values of the delay t_1 gives rise to a *modulated* series of spectra after Fourier transformation (Fig. 2.5(b)). If the column of data points labelled C in Fig. 2.5(b), corresponding to the centre of the peak, is taken from the modulated data set, then it will look exactly like a free induction decay except that its time axis is not t_2 but t_1 (Fig. 2.5(c)).

This *interferogram* can be Fourier transformed in the usual way to give the normal NMR spectrum once again (Fig. 2.5(d)). The Fourier transform of column B gives the same spectrum in this case, but with lower amplitude. By now it should be clear that a full two-dimensional spectrum can be arrived at by Fourier transforming *all* the columns of data points in Fig. 2.5(b). More formally, the result of performing a two-dimensional experiment is a time-domain data set $s(t_1,t_2)$; the free induction decays are all Fourier transformed to give a series of modulated spectra $S(t_1,F_2)$; and finally the interferograms are Fourier transformed to give a two-dimensional spectrum $S(F_1,F_2)$. When performing the two-dimensional

Here *t* is used to indicate time and *F* is used to indicate frequency; the notation is traditional.

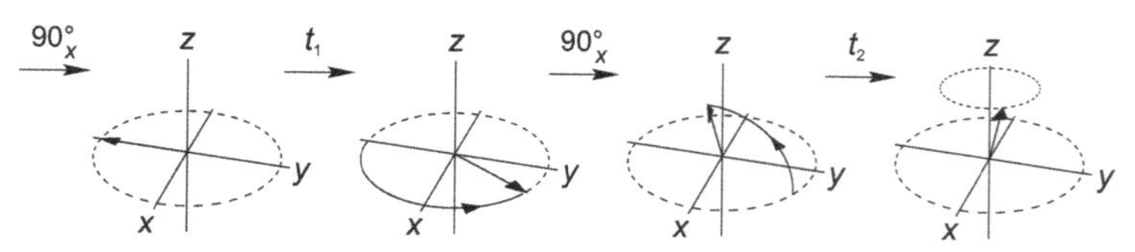

Fig. 2.4 Evolution of the magnetization vector during the pulse sequence of Fig. 2.3. The initial amplitude of the free induction decay at $t_2 = 0$ is determined by Ωt_1, the angle through which $\boldsymbol{M}$ precesses during t_1.

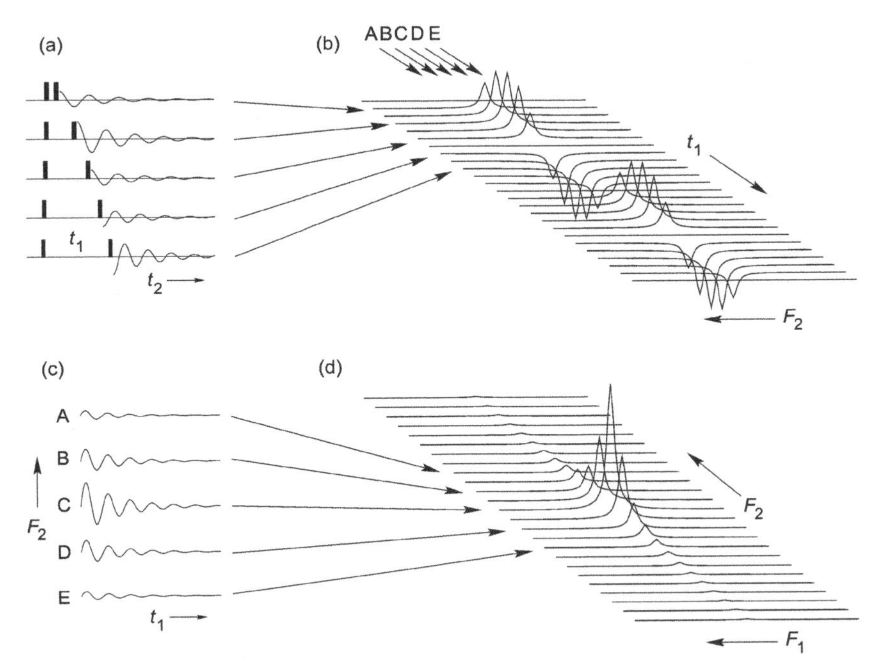

Fig. 2.5 The operation of a two-dimensional NMR experiment (Fig. 2.3) on a sample with a single NMR line. (a) Free induction decays for different values of the inter-pulse delay t_1. (b) The spectra obtained by Fourier transforming several such signals. (c) Interferograms constructed by extracting columns of data points from (b) at the indicated positions (A–E). (d) The two-dimensional spectrum that results from Fourier transformation of the interferograms. Only a few of the free induction decays and interferograms are shown for clarity.

experiment, the magnetizations evolving during t_1 must be sampled regularly according to the same constraints as the data points in t_2. If a spectral width of *SW* hertz is required (i.e. *SW*/2 either side of the transmitter frequency), the interval between t_1 (or t_2) samples should be 1/*SW* seconds, assuming quadrature detection is used.

The type of modulation described above is known as *amplitude modulation*. The free induction decays observed in t_2 are modulated in amplitude as a function of t_1 but their initial phases are always the same. The majority of important two-dimensional experiments (COSY, NOESY, etc.) inherently yield amplitude-modulated data which has the advantage that the spectra are easily processed to give pure absorption two-dimensional lineshapes. The disadvantage is that the sense of the precession of the magnetization during t_1 is not determined (the same problem as experienced when using a single channel detector in one-dimensional NMR). Special methods have been developed to solve this problem (see Section 2.4). Experiments that do not possess a mixing period generally yield *phase-modulated* data as a function of t_1. The best known examples are some NMR imaging experiments and the class of two-dimensional techniques known as *J spectroscopy*, neither of which will be discussed here.

Experiments such as COSY and NOESY will be discussed in Chapter 5.

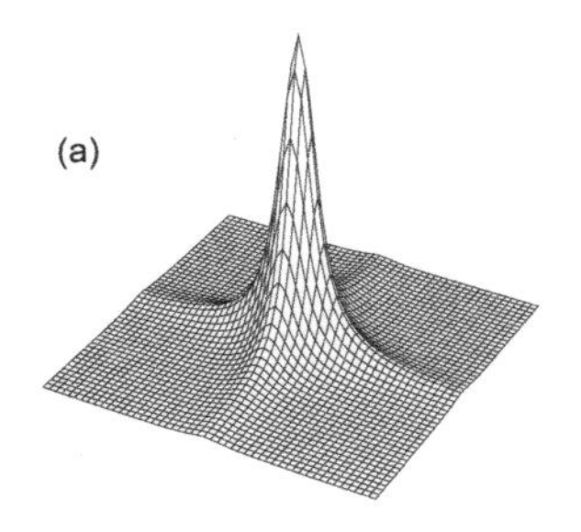

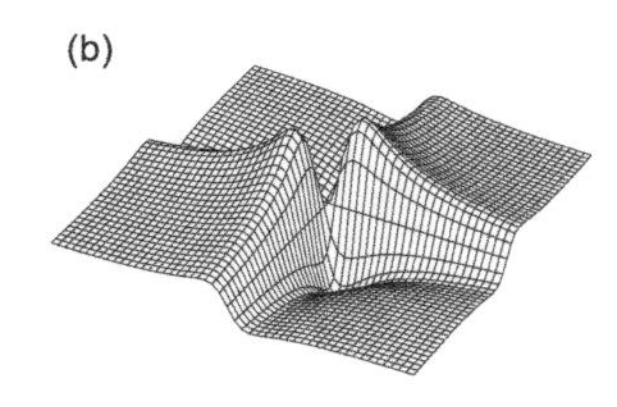

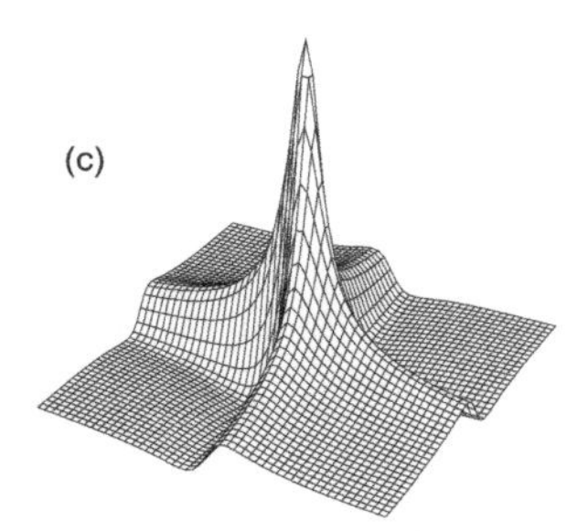

Fig. 2.6 Two-dimensional lineshapes: (a) absorption in both dimensions; (b) dispersion in both dimensions; (c) the phase twist lineshape, a combination of (a) and (b), whose broad wings and negative regions make it highly undesirable.

2.4 Pure phase two-dimensional spectra

The simple two-dimensional experiment discussed in Section 2.3 gives data of the general form:

$$s(t_1,t_2)=\cos\Omega t_1\exp(-t_1/T_2)\exp(\mathrm{i}\Omega t_2)\exp(-t_2/T_2) \tag{2.7}$$

where T_2 is the spin-spin relaxation time. In Section 2.2, we saw that the complex Fourier transform of a signal of the type $\exp(\mathrm{i}\Omega t)\exp(-t/T_2)$ is $A(\Delta\omega)-\mathrm{i}D(\Delta\omega)$ with $\Delta\omega=\omega-\Omega$. We will here abbreviate the notation further by writing $A_k^{\pm}$ and $D_k^{\pm}$ for absorptive and dispersive lineshapes centred at frequency $\pm\Omega$ in the F_k dimension. Complex Fourier transformation of $\exp(\pm\mathrm{i}\Omega t_k)\exp(-t_k/T_2)$ thus gives $A_k^{\pm}-\mathrm{i}D_k^{\pm}$. The Fourier transform of $s(t_1,t_2)$ with respect to t_2 is, therefore,

$$\begin{aligned}S(t_1,F_2)&=\cos\Omega t_1\exp(-t_1/T_2)\left[A_2^+-\mathrm{i}D_2^+\right]\\&=\tfrac{1}{2}\left[\exp(+\mathrm{i}\Omega t_1)+\exp(-\mathrm{i}\Omega t_1)\right]\exp(-t_1/T_2)\left[A_2^+-\mathrm{i}D_2^+\right].\end{aligned} \tag{2.8}$$

A complex Fourier transform of this with respect to t_1 gives

$$\begin{aligned}S(F_1,F_2)&=\tfrac{1}{2}\left[A_1^+-\mathrm{i}D_1^++A_1^--\mathrm{i}D_1^-\right]\times\left[A_2^+-\mathrm{i}D_2^+\right]\\&=\tfrac{1}{2}\left[\left(A_1^+A_2^+-D_1^+D_2^+\right)+\left(A_1^-A_2^+-D_1^-D_2^+\right)\right.\\&\quad\left.-\mathrm{i}\left(A_1^+D_2^++D_1^+A_2^+\right)-\mathrm{i}\left(A_1^-D_2^++D_1^-A_2^+\right)\right].\end{aligned} \tag{2.9}$$

The mixture of two-dimensional absorption and dispersion lineshapes represented by terms such as $\left(A_1^+A_2^+-D_1^+D_2^+\right)$ is the dreaded *phase twist* lineshape (Fig. 2.6). The real part of the two-dimensional spectrum described by Eqn 2.9 is thus a phase twist positioned at $(F_1,F_2)=(+\Omega,+\Omega)$ and a phase twist at $(F_1,F_2)=(-\Omega,+\Omega)$. This is the worst of both worlds: we have obtained phase twist lineshapes and we have not yet achieved quadrature detection in F_1 and so cannot distinguish positive and negative frequencies in F_1. Clearly we are doing something wrong.

The answer is to record and store separately two complete two-dimensional time-domain signals, one with cosine modulation as above, the other with sine modulation with respect to t_1. The latter is obtained by repeating the whole experiment with different pulse phases. It is also important to distinguish carefully between imaginary data in the two frequency dimensions, which can be achieved by writing the imaginary unit as i_1 when used for t_1 and as i_2 when used for t_2. Thus we have

Note that i_1 and i_2 are not really two different square roots of -1, but simply a notational convenience. This data-processing trick of keeping the imaginary parts of the signal separated in the two frequency dimensions is often referred to as a *hypercomplex Fourier transformation*.

$$\begin{aligned}s_{\cos}(t_1,t_2)&=\cos(\Omega t_1)\exp(-t_1/T_2)\exp(\mathrm{i}_2\Omega t_2)\exp(-t_2/T_2)\\s_{\sin}(t_1,t_2)&=\sin(\Omega t_1)\exp(-t_1/T_2)\exp(\mathrm{i}_2\Omega t_2)\exp(-t_2/T_2).\end{aligned} \tag{2.10}$$

Fourier transformation of these two signals with respect to t_2 gives

$$\begin{aligned}S_{\cos}(t_1,F_2)&=\cos(\Omega t_1)\exp(-t_1/T_2)\left[A_2^+-\mathrm{i}_2D_2^+\right]\\S_{\sin}(t_1,F_2)&=\sin(\Omega t_1)\exp(-t_1/T_2)\left[A_2^+-\mathrm{i}_2D_2^+\right].\end{aligned} \tag{2.11}$$

The signal S_{sin} is now multiplied by the imaginary unit i_1 and added to S_{cos}

$$S_{cos}(t_1,F_2)+i_1 S_{sin}(t_1,F_2)=\exp(i_1\Omega t_1)\exp(-t_1/T_2)\left[A_2^+ - i_2 D_2^+\right]. \quad (2.12)$$

Finally, Fourier transformation with respect to t_1 gives a spectrum made up of four separate components

$$\begin{aligned} S(F_1,F_2) &= \left[A_1^+ - i_1 D_1^+\right]\times\left[A_2^+ - i_2 D_2^+\right] \\ &= A_1^+ A_2^+ - i_1 D_1^+ A_2^+ - i_2 A_1^+ D_2^+ + i_1 i_2 D_1^+ D_2^+ . \end{aligned} \quad (2.13)$$

As desired, we have achieved quadrature detection in F_1 and the real part represents a two-dimensional absorption lineshape (see Fig. 2.6) at $(F_1,F_2)=(+\Omega,+\Omega)$.

2.5 Decoupling

Decoupling is an experimental technique used to remove the effects of unwanted J-couplings in a spectrum. In simple NMR spectra, couplings usually provide useful information about molecular structure, but in two-dimensional NMR such information can sometimes be obtained more simply from the presence of *cross peaks*, and so it may be useful to remove couplings from the F_1 or F_2 dimension.

Cross peaks will be discussed in Chapter 5.

The basic method is a generalization of the heteronuclear spin echo described in Section 1.7. Consider again the spectrum of a ^1H nucleus coupled to a ^{13}C, for which the ^1H NMR spectrum is a doublet, centred at the ^1H resonance offset Ω with splitting $2\pi J$. As described in Section 2.2, the ^1H free induction decay is not observed continuously, but sampled at a series of discrete time points, separated by constant intervals Δt. If these time intervals are long enough then it is possible to apply a 180° pulse to the ^{13}C nucleus at the midpoint of each interval, completely refocusing any evolution under the J-coupling. Thus, the final ^1H NMR spectrum is a singlet, centred at the ^1H resonance offset Ω. Note that the ^1H chemical shift evolution is not refocused, as the 180° pulses are only applied to the ^{13}C nuclei.

Treating decoupling as a series of 180° pulses synchronized with the digitization of the FID is the simplest way to think about the process, but not a good way to implement it. The simplest approach to decoupling is to apply a *continuous* strong radiofrequency field to the partner nucleus throughout acquisition. The best approach requires that the phase of the decoupling field follows a complex pattern of shifts during the decoupling period, a technique usually called composite pulse decoupling.

There are many composite pulse decoupling sequences optimized for different experimental conditions, but the subject is too complex to consider here. For more details see Freeman *Spin Choreography* (1998) and Shaka and Keeler *Broadband Spin Decoupling in Isotropic Liquids* (1987).

2.6 Summary

- NMR spectrometers detect precessing transverse magnetization, known as the free induction decay.
- The frequencies present in the free induction decay are the precession frequencies in the rotating frame.

- The spectrum is the Fourier transform of the free induction decay.
- Quadrature detection is used to distinguish positive and negative precession frequencies.
- Absorptive lineshapes are preferable to dispersive lineshapes.
- It is essential to 'phase the spectrum' to obtain absorptive lineshapes.
- The interval between points in a free induction decay is the reciprocal of the spectral width.
- Two-dimensional NMR involves two distinct periods of free precession separated by a mixing period.
- Quadrature detection and pure absorptive lineshapes in two-dimensional NMR spectra require multiple datasets to be acquired.
- Decoupling can be used to remove heteronuclear J-couplings if desired.

2.7 Exercises

Worked solutions to the exercises are available on the Online Resource Centre at www.oxfordtextbooks.co.uk/orc/hore2e/

2.1 Verify that Eqns 2.1 and 2.2 are consistent with Eqns 2.3 and 2.4.

2.2 Verify that the full width (in Hz) at half maximum height of the absorption lineshape $A(\Delta\omega)$ in Eqn 2.4 is $1/\pi T_2$.

2.3 Plot $A(\Delta\omega)$ and $D(\Delta\omega)$ (Eqn 2.4) to show that $D(\Delta\omega)$ has a lot more intensity in its wings than does $A(\Delta\omega)$. Take $T_2 = 1$ s and $-20 \leq \Delta\omega \leq +20$ rad s^{-1}.

2.4 Determine the appearance of the spectrum corresponding to the model free induction decay: $\left[\exp(i\Omega_A t) + i\exp(i\Omega_B t)\right]\exp(-t/T_2)$.

2.5 Determine the appearance of the spectra corresponding to the model free induction decays: (a) $\cos(\pi Jt)\exp(i\Omega t)\exp(-t/T_2)$ and (b) $\cos^2(\pi Jt)\exp(i\Omega t)\exp(-t/T_2)$.

2.6 This exercise is intended to verify that phase-correction works, as depicted in Fig. 2.2. Consider a free induction decay of the general form $s(t) = \exp(i\Omega t)\exp(i\phi)\exp(-t/T_2)$. Show that the spectrum $R\cos\phi + I\sin\phi$ is an absorption lineshape. R and I are the real and imaginary parts of the Fourier transform of $s(t)$.

2.7 Show that phase-modulated two-dimensional signals of the general form $s(t_1,t_2) = \exp(i\Omega t_1)\exp(-t_1/T_2)\exp(i\Omega t_2)\exp(-t_2/T_2)$ allow quadrature detection in both dimensions but do not give pure absorptive lineshapes.

2.8 A common error in many real spectrometers is that analogue-to-digital converters are not perfectly balanced. As a consequence, their outputs often contain a constant offset in addition to the free induction signal. What effect will this have on the NMR spectrum obtained by Fourier transformation?

2.9 A second common problem with quadrature detection is that the two detection channels have slightly different sensitivities so that, for example, the same signal would give a slightly bigger output in the real channel than in the imaginary channel. What effect will this have on the NMR spectrum?

Product operators I

3.1 Introduction

Product operators became popular as a way of describing NMR experiments after the publication of a review article by Ernst's group. As the word 'operator' indicates, they are inherently quantum-mechanical in nature; however, they can be successfully used with no quantum-mechanical knowledge whatsoever. All that must be learned are a few simple rules, although some familiarity with basic trigonometry is useful. The great beauty of product operators is that they largely retain the geometrical picture and intuitive 'feel' of the vector model, while offering an exact analysis (under conditions of weak J-coupling) of the inner workings of modern NMR experiments.

See Sørensen *et al. Product Operator Formalism for the Description of NMR Pulse Experiments* (1983).

A pair of spins is said to be *weakly coupled* when its spin-spin coupling constant J is small compared with the difference in resonance frequencies due to chemical shifts.

3.2 Product operators for one spin

Four operators suffice to describe NMR experiments on molecules containing an isolated spin-$\frac{1}{2}$ nucleus. For a spin labelled I they are:

$$\tfrac{1}{2}E \quad I_x \quad I_y \quad I_z. \tag{3.1}$$

The first is simply half the identity operator, and is included for purely formal reasons. It is the remaining three operators that will concern us. They correspond to the x-, y-, and z-magnetization of a single spin, in the rotating frame (Fig. 3.1).

We use product operators here not as operators as such but as a way of describing both the state of the spin system, e.g. before and after a pulse, and the interactions which cause the state to evolve. We use the term 'delay' as an NMR colloquialism for an interval of free precession. As we shall see in Part B, formal operators are needed to *calculate* the effects of pulses and delays. Here we simply provide 'recipes' that enable one to do these transformations. The link between operators and states will also become clear in Part B.

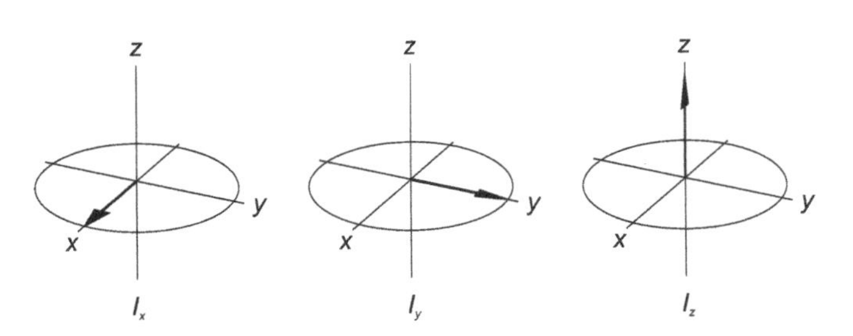

Fig. 3.1 Vector representation of the product operators I_x, I_y, and I_z for a single spin-$\frac{1}{2}$ nucleus.

From now on we describe all evolutions by listing the operator corresponding to the evolution (e.g. I_x) and the extent of this evolution, which can be described by an angle (here 90°). Thus a 90°_x pulse applied to spin I is designated by $90°I_x$.

From the vector model, it is easy to see how I_x, I_y, and I_z transform under (say) a 90° pulse.

$$
\begin{aligned}
&I_x \xrightarrow{90°I_x} I_x \qquad && I_x \xrightarrow{90°I_y} -I_z \\
&I_y \xrightarrow{90°I_x} I_z \qquad && I_y \xrightarrow{90°I_y} I_y \\
&I_z \xrightarrow{90°I_x} -I_y \qquad && I_z \xrightarrow{90°I_y} I_x.
\end{aligned}
\tag{3.2}
$$

More generally, the effect of pulses with flip angle β on the three operators is given by:

$$
\begin{aligned}
&I_x \xrightarrow{\beta I_x} I_x \qquad && I_x \xrightarrow{\beta I_y} I_x \cos\beta - I_z \sin\beta \\
&I_y \xrightarrow{\beta I_x} I_y \cos\beta + I_z \sin\beta \qquad && I_y \xrightarrow{\beta I_y} I_y \\
&I_z \xrightarrow{\beta I_x} I_z \cos\beta - I_y \sin\beta \qquad && I_z \xrightarrow{\beta I_y} I_z \cos\beta + I_x \sin\beta.
\end{aligned}
\tag{3.3}
$$

For a single spin, free precession around the z-axis during a delay is described by evolution under the I_z operator; the evolution angle depends on the precession frequency Ω and the evolution time t.

Similarly, it is easy to see from the vector model how the three operators transform as a result of free precession for a time t at a resonance offset Ω:

$$
\begin{aligned}
&I_x \xrightarrow{\Omega t I_z} I_x \cos\Omega t + I_y \sin\Omega t \\
&I_y \xrightarrow{\Omega t I_z} I_y \cos\Omega t - I_x \sin\Omega t \\
&I_z \xrightarrow{\Omega t I_z} I_z.
\end{aligned}
\tag{3.4}
$$

As a reminder of the sign conventions involved in these rotations, which are the same as those we have used for the vector model, we draw the hybrid vector model/product operator diagrams shown in Fig. 3.2.

Product operators can also be related to spectra and energy-level diagrams, as indicated in Fig. 3.3 for a spin-$\frac{1}{2}$ nucleus. The dotted lines indicate the 90° phase

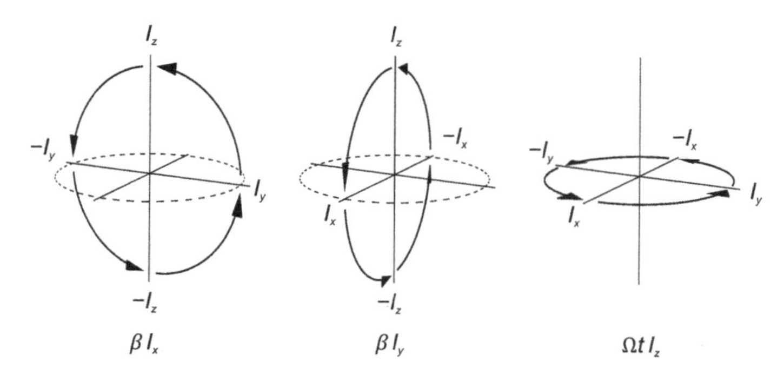

Fig. 3.2 The transformations of the product operators of a single spin-$\frac{1}{2}$ nucleus produced by x or y pulses (flip angle β), or a period t of free precession at a resonance offset Ω. These diagrams are nothing more than a pictorial representation of Eqns 3.3 and 3.4. For example, a βI_y pulse transforms I_x into $I_x\cos\beta - I_z\sin\beta$, and $-I_z$ into $-I_z\cos\beta - I_x\sin\beta$.

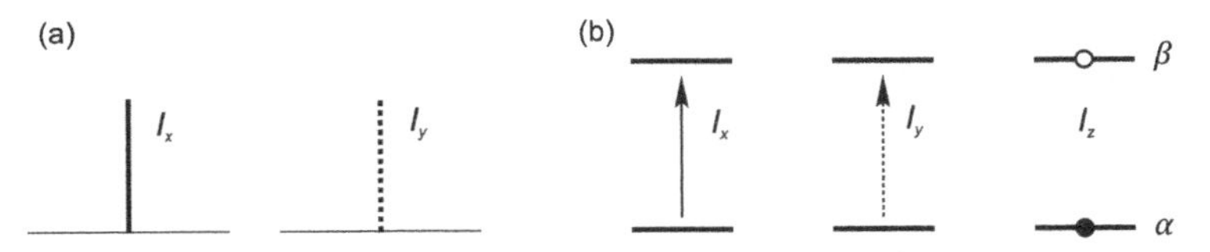

Fig. 3.3 Representation of the product operators for a single spin-$\frac{1}{2}$ nucleus in terms of (a) schematic spectra and (b) energy-level diagrams. The dotted line for I_y is a reminder of the 90° phase difference from I_x. The arrows correspond to the transitions with which product operators are associated. The open and closed circles indicate, respectively, a population deficit and a population excess in the corresponding energy level, as would be found at thermal equilibrium for example.

difference between I_x and I_y operators. Thus, if we have two spins I and S with the state of the system being described by $I_x + S_y$, then if the I peak is phased to pure absorption the S peak will be in dispersion (or vice versa).

To illustrate the use of product operators, we can look at the effect of the spin echo sequence $90^\circ_x - \tau - 180^\circ_y - \tau$ on a molecule with a one-line NMR spectrum. An echo is formed because the two free precession times either side of the 180° pulse are equal. At each stage in this 'calculation' the rules in Eqns 3.3 and 3.4 are applied to each of the product operators in turn, starting with the thermal equilibrium state I_z:

$$
\begin{aligned}
I_z &\xrightarrow{90^\circ I_x} -I_y \\
&\xrightarrow{\Omega\tau I_z} -I_y\cos\Omega\tau + I_x\sin\Omega\tau \\
&\xrightarrow{180^\circ I_y} -I_y\cos\Omega\tau - I_x\sin\Omega\tau \\
&\xrightarrow{\Omega\tau I_z} -\left(I_y\cos\Omega\tau - I_x\sin\Omega\tau\right)\cos\Omega\tau \\
&\qquad -\left(I_x\cos\Omega\tau + I_y\sin\Omega\tau\right)\sin\Omega\tau \\
&= \quad -I_y\cos^2\Omega\tau + I_x\sin\Omega\tau\cos\Omega\tau \\
&\qquad -I_x\cos\Omega\tau\sin\Omega\tau - I_y\sin^2\Omega\tau. \qquad (3.5)
\end{aligned}
$$

The I_x terms cancel and we are left with:

$$-I_y\cos^2\Omega\tau - I_y\sin^2\Omega\tau = -I_y, \qquad (3.6)$$

N.B. $\cos^2 A + \sin^2 A = 1$

i.e. the same state as was created immediately after the initial 90°_x pulse.

All this may seem a bit of a cheat, however; although the product operators give the same result as the vector model (Fig. 1.6), this is hardly surprising given that we have simply used them as a sort of 'vector model with numbers on it'. It is when we consider systems of *J*-coupled nuclei that product operators come into their own and we can start to explain experiments that cannot be understood using the vector model. Although the vector model can be extended to describe simple experiments with *J*-coupled nuclei it cannot be used to describe most modern NMR experiments.

3.3 Product operators for two coupled spins

The restriction to weak coupling is the one serious limitation of the product operator approach. We deal with strong coupling in Part B.

We consider the NMR spectrum of two *weakly J*-coupled spin-$\frac{1}{2}$ nuclei, labelled I and S.

The schematic spectrum in Fig. 3.4 may be taken to represent a homonuclear spin system, where I and S are both (say) protons, or a heteronuclear spin system (e.g. where I is ^{1}H and S is ^{13}C) in which case S is typically several hundred megahertz off resonance and is not affected by any pulses applied to the I spin. As the name implies, the 16 product operators for a system of two spin-$\frac{1}{2}$ nuclei can be formed by taking the products of the four operators for the individual spins I and S:

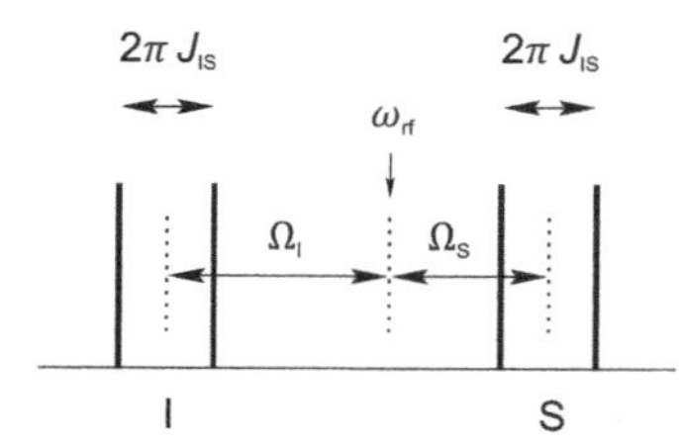

Fig. 3.4 Schematic NMR spectrum of a pair of spin-$\frac{1}{2}$ nuclei, I and S. Here ω_{rf} is the transmitter frequency, Ω_I and Ω_S are the offset frequencies, and J_{IS} the spin-spin coupling constant. Note that ω_{rf}, Ω_I, Ω_S, and $2\pi J_{IS}$ are all angular frequencies.

$$2\times\quad\begin{array}{c|cccc} & \frac{1}{2}E & S_x & S_y & S_z \\ \hline \frac{1}{2}E & \frac{1}{2}E & S_x & S_y & S_z \\ I_x & I_x & 2I_xS_x & 2I_xS_y & 2I_xS_z \\ I_y & I_y & 2I_yS_x & 2I_yS_y & 2I_yS_z \\ I_z & I_z & 2I_zS_x & 2I_zS_y & 2I_zS_z \end{array} \tag{3.7}$$

The factor of two on the left is one of those normalizations beloved of theoreticians. For the moment we will ignore the four product operators in the centre of the table ($2I_xS_x$, $2I_xS_y$, $2I_yS_x$, $2I_yS_y$). It will be shown later (Section 4.3) that these represent *multiple-quantum coherences*. The remaining product operators can be related to the vector model, as shown in Fig. 3.5. I_x, I_y, I_z, S_x, S_y, and S_z are referred to as *in-phase* operators, while $2I_xS_z$, $2I_yS_z$, $2I_zS_x$, $2I_zS_y$, and $2I_zS_z$ are *antiphase* operators. Once again, the product operators can be related to schematic NMR spectra and energy-level diagrams (Fig. 3.6).

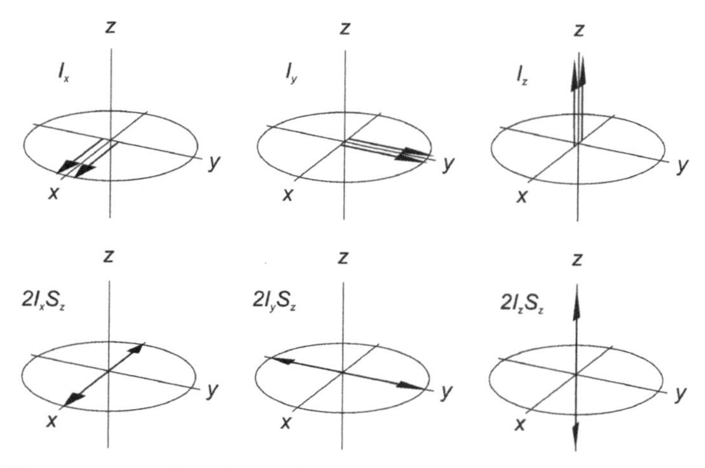

Fig. 3.5 Vector representation of some of the product operators for a two-spin IS system.

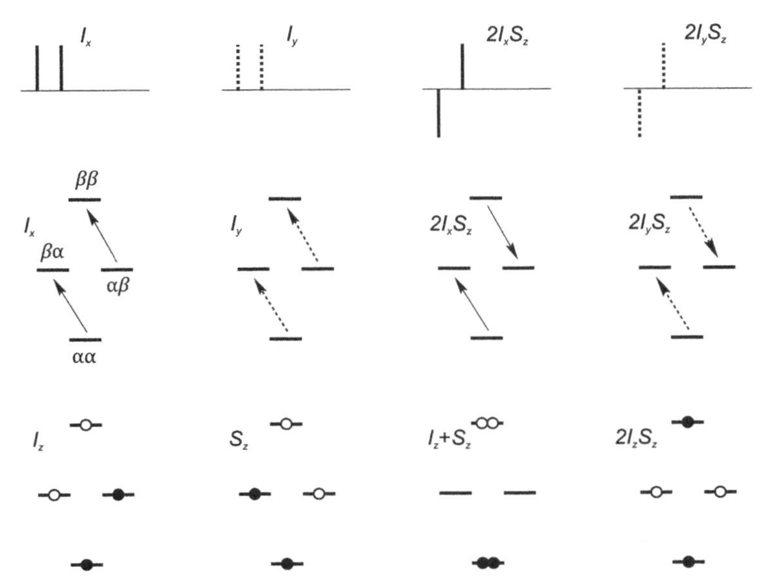

Fig. 3.6 Representation of some of the product operators for a two-spin IS system in terms of schematic spectra (above) and energy-level diagrams (below). The corresponding S-spin operators have the same form, centred on Ω_S. Dotted lines, arrows, and open/closed circles have the same significance as in Fig. 3.3.

The operator corresponding to evolution under a *J*-coupling is $2I_zS_z$, and the *new* transformation rules, summarized in Fig. 3.7, for evolution under a *J*-coupling J_{IS} for a period *t* are:

We will see later where the πJ_{IS} frequency in these expressions comes from. For now, note that they are reassuringly consistent with the vector model (Figs. 1.7 and 1.8). The choice of the operator $2I_zS_z$ to describe the evolution is discussed in Part B.

$$
\begin{aligned}
I_x &\xrightarrow{\pi J_{IS}t2I_zS_z} I_x\cos(\pi J_{IS}t)+2I_yS_z\sin(\pi J_{IS}t)\\
I_y &\xrightarrow{\pi J_{IS}t2I_zS_z} I_y\cos(\pi J_{IS}t)-2I_xS_z\sin(\pi J_{IS}t)\\
I_z &\xrightarrow{\pi J_{IS}t2I_zS_z} I_z\\
2I_xS_z &\xrightarrow{\pi J_{IS}t2I_zS_z} 2I_xS_z\cos(\pi J_{IS}t)+I_y\sin(\pi J_{IS}t)\\
2I_yS_z &\xrightarrow{\pi J_{IS}t2I_zS_z} 2I_yS_z\cos(\pi J_{IS}t)-I_x\sin(\pi J_{IS}t)\\
2I_zS_z &\xrightarrow{\pi J_{IS}t2I_zS_z} 2I_zS_z
\end{aligned}
\qquad (3.8)
$$

and similarly for S_x, S_y, S_z, $2I_zS_x$, and $2I_zS_y$.

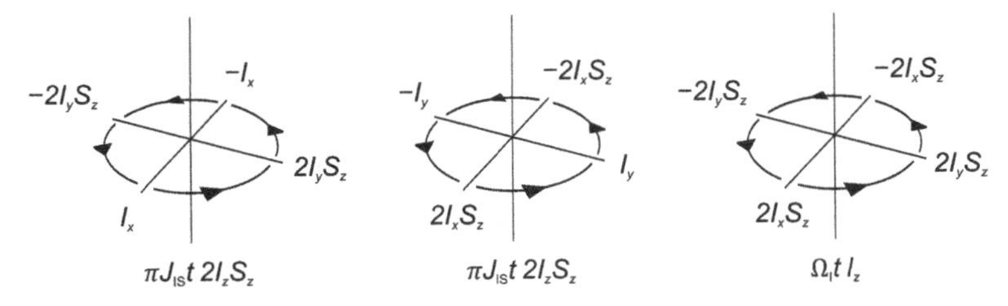

Fig. 3.7 The transformations of the product operators for a two-spin IS system produced by the scalar coupling J_{IS} and the resonance offset Ω_I during a period *t*.

The transformation rules for pulses and free precession under the resonance offset follow immediately from Eqns 3.3 and 3.4. For example:

Note that I-spin operators only affect the state of spin I, and similarly for spin S.

$$2I_xS_z \xrightarrow{\Omega_I t I_z} 2I_xS_z\cos\Omega_I t + 2I_yS_z\sin\Omega_I t$$
$$2I_zS_y \xrightarrow{\Omega_S t S_z} 2I_zS_y\cos\Omega_S t - 2I_zS_x\sin\Omega_S t$$
$$2I_xS_z \xrightarrow{\beta I_y} 2I_xS_z\cos\beta - 2I_zS_z\sin\beta. \quad (3.9)$$

In general, however, both spins can evolve at the same time: for example, during a delay each spin will evolve at its own offset frequency. As each evolution only affects spin I or spin S the two evolutions can be calculated sequentially in any order

Remember that the S_z state does not evolve under S_z, i.e. z-magnetization does not precess during delays; see Eqn 3.4.

$$2I_xS_z \xrightarrow{\Omega_I t I_z} 2I_xS_z\cos\Omega_I t + 2I_yS_z\sin\Omega_I t$$
$$\xrightarrow{\Omega_S t S_z} 2I_xS_z\cos\Omega_I t + 2I_yS_z\sin\Omega_I t. \quad (3.10)$$

The same approach can be used if a pulse is applied simultaneously to spins I and S, but in this case it is usual to write the combined evolution operator as the sum of the two individual terms and then carry out the two calculations in one step:

$$2I_zS_y \xrightarrow{90^\circ(I_x+S_x)} -2I_yS_z. \quad (3.11)$$

All this may seem pretty abstract and of questionable value. So let us analyse in some detail a couple of simple examples where we know what the answer should be from the vector model.

3.4 Spin echoes

The product operator description of a spin echo in a J-coupled two-spin IS system proceeds as follows (the algebra is not as frightful as it looks). We begin with a *heteronuclear* IS spin system where the 90° and 180° pulses only affect the spin I, and concentrating on this spin we have:

Once again, although the evolutions caused by the chemical shift (Ω_I) and the spin-spin coupling (πJ_{IS}) actually occur simultaneously, we can treat them as if they were sequential, and the order in which we apply them is immaterial; this will be justified in Part B. The S spin is unaffected by all these evolutions, and so begins and ends in the state S_z.

$$I_z \xrightarrow{90^\circ I_x} -I_y$$
$$\xrightarrow{\Omega_I \tau I_z} -I_y\cos\Omega_I\tau + I_x\sin\Omega_I\tau$$
$$\xrightarrow{\pi J_{IS}\tau 2I_zS_z} -I_y\cos\Omega_I\tau\cos(\pi J_{IS}\tau) + 2I_xS_z\cos\Omega_I\tau\sin(\pi J_{IS}\tau)$$
$$+I_x\sin\Omega_I\tau\cos(\pi J_{IS}\tau) + 2I_yS_z\sin\Omega_I\tau\sin(\pi J_{IS}\tau). \quad (3.12)$$

The 180° pulse only affects the I spin operators and we can write

$$\xrightarrow{180^\circ I_y} -I_y\cos\Omega_I\tau\cos(\pi J_{IS}\tau) - 2I_xS_z\cos\Omega_I\tau\sin(\pi J_{IS}\tau)$$
$$-I_x\sin\Omega_I\tau\cos(\pi J_{IS}\tau) + 2I_yS_z\sin\Omega_I\tau\sin(\pi J_{IS}\tau). \quad (3.13)$$

Now for the chemical shift evolution during the second free precession period:

$$\xrightarrow{\Omega_I \tau I_z} -\left(I_y \cos\Omega_I\tau - I_x \sin\Omega_I\tau\right)\cos\Omega_I\tau \cos(\pi J_{IS}\tau)$$
$$-\left(2I_xS_z \cos\Omega_I\tau + 2I_yS_z \sin\Omega_I\tau\right)\cos\Omega_I\tau \sin(\pi J_{IS}\tau)$$
$$-\left(I_x \cos\Omega_I\tau + I_y \sin\Omega_I\tau\right)\sin\Omega_I\tau \cos(\pi J_{IS}\tau)$$
$$+\left(2I_yS_z \cos\Omega_I\tau - 2I_xS_z \sin\Omega_I\tau\right)\sin\Omega_I\tau \sin(\pi J_{IS}\tau). \quad (3.14)$$

This looks horrible, but the terms in I_x cancel, as do those containing $2I_yS_z$, to give

$$-I_y \cos^2\Omega_I\tau \cos(\pi J_{IS}\tau) - I_y \sin^2\Omega_I\tau \cos(\pi J_{IS}\tau)$$
$$-2I_xS_z \cos^2\Omega_I\tau \sin(\pi J_{IS}\tau) - 2I_xS_z \sin^2\Omega_I\tau \sin(\pi J_{IS}\tau)$$
$$= -I_y \cos(\pi J_{IS}\tau) - 2I_xS_z \sin(\pi J_{IS}\tau). \quad (3.15)$$

Note that all terms involving $\Omega\tau$ have disappeared. This is as expected, of course: the spin echo has refocused the chemical shift. To complete the calculation there must be evolution under the coupling in the second τ period:

$$\xrightarrow{\pi J_{IS}\tau 2I_zS_z} -I_y \cos^2(\pi J_{IS}\tau) + 2I_xS_z \cos(\pi J_{IS}\tau)\sin(\pi J_{IS}\tau)$$
$$-2I_xS_z \sin(\pi J_{IS}\tau)\cos(\pi J_{IS}\tau) - I_y \sin^2(\pi J_{IS}\tau)$$
$$= -I_y. \quad (3.16)$$

As predicted by the vector model, both the J-coupling and the chemical shift have been refocused for the heteronuclear case.

The analysis of a *homonuclear* spin system is identical up to the 180°_y pulse, which now changes the sign of all x and z operators of *both* spins:

The first stage is, of course, exactly the same as Eqn 3.12. Note that $2I_xS_z$ changes sign *twice*, and so retains its original sign.

$$I_z \xrightarrow{90^\circ I_x} \xrightarrow{\Omega_I \tau I_z} \xrightarrow{\pi J_{IS}\tau 2I_zS_z}$$
$$-I_y \cos\Omega_I\tau \cos(\pi J_{IS}\tau) + 2I_xS_z \cos\Omega_I\tau \sin(\pi J_{IS}\tau)$$
$$+I_x \sin\Omega_I\tau \cos(\pi J_{IS}\tau) + 2I_yS_z \sin\Omega_I\tau \sin(\pi J_{IS}\tau)$$
$$\xrightarrow{180^\circ(I_y+S_y)} -I_y \cos\Omega_I\tau \cos(\pi J_{IS}\tau) + 2I_xS_z \cos\Omega_I\tau \sin(\pi J_{IS}\tau)$$
$$-I_x \sin\Omega_I\tau \cos(\pi J_{IS}\tau) - 2I_yS_z \sin\Omega_I\tau \sin(\pi J_{IS}\tau). \quad (3.17)$$

Now for the chemical shift evolution during the second free precession period:

$$\xrightarrow{\Omega_I \tau I_z} -I_y \cos(\pi J_{IS}\tau) + 2I_xS_z \sin(\pi J_{IS}\tau). \quad (3.18)$$

In this step we have implicitly carried out the cancellations and trigonometric reduction which reveal that the chemical shift has been refocused. This expression differs from the heteronuclear case in Eqn 3.15 only in the sign of the antiphase term $2I_xS_z$. Finally, the evolution under the coupling in the second τ period can be considered:

N.B. $\cos^2 A - \sin^2 A = \cos 2A$ and $2\sin A\cos A = \sin 2A$

$$\xrightarrow{\pi J_{IS}\tau 2I_zS_z} -I_y \cos^2(\pi J_{IS}\tau) + 2I_xS_z \cos(\pi J_{IS}\tau)\sin(\pi J_{IS}\tau)$$
$$+2I_xS_z \sin(\pi J_{IS}\tau)\cos(\pi J_{IS}\tau) + I_y \sin^2(\pi J_{IS}\tau)$$
$$= -I_y \cos(2\pi J_{IS}\tau) + 2I_xS_z \sin(2\pi J_{IS}\tau). \quad (3.19)$$

Reassuringly, this is exactly the result predicted by the vector model (Fig. 1.8). Only the J-coupling has evolved, at a frequency πJ_{IS} for a time 2τ. Remembering that in this homonuclear spin system I behaves exactly the same as S, the full result becomes:

$$-(I_y + S_y)\cos(2\pi J_{IS}\tau) + (2I_xS_z + 2I_zS_x)\sin(2\pi J_{IS}\tau). \tag{3.20}$$

3.5 Summary

- For isolated spins, the product operator formalism is just an algebraic version of the geometric vector model.
- For pairs of coupled spins, product operators enable a simple but mathematically correct treatment of pulses and delays provided that the spins are weakly coupled.
- Product operators can be related to schematic spectra and energy levels.

3.6 Exercises

Worked solutions to the exercises are available on the Online Resource Centre at www.oxfordtextbooks.co.uk/orc/hore2e/

3.1 Show that the pulse sequence $90^\circ_x - \tau - 90^\circ_{-x}$ has the following result: $I_z \longrightarrow I_z \cos\Omega\tau + I_x \sin\Omega\tau$. A sample comprises a dilute solute in an otherwise pure solvent; the solute and solvent each give an NMR singlet. How might this pulse sequence be used to obtain a spectrum of the solute free from the obscuring presence of the solvent line?

3.2 Determine the effect of the 'composite pulse' $90^\circ_{-x} - \theta_y - 90^\circ_x$ on the initial states: (a) I_x, (b) I_y, and (c) I_z. What geometric operation does this composite pulse perform?

3.3 The evolution of the one-spin operators described in Eqns 3.3 and 3.4 and in Fig. 3.2 can also be summarized in a 3×3 table. Using the three initial states (I_x, I_y, and I_z) to label the rows and the operators corresponding to the three basic transformations (I_x, I_y, and I_z) to label the columns, enter into the table the result *towards which each operator evolves*. For example, $I_y \xrightarrow{\beta I_x} I_y \cos\beta + I_z \sin\beta$ (Eqn 3.3) so that the entry in the I_y-row and I_x-column would be I_z. If an operator does not evolve then enter 0 for the corresponding element of the table.

3.4 Repeat Exercise 3.3 for two-spin operators by forming a table for the 11 initial states (I_q, S_q, $2I_zS_q$, $2I_qS_z$, q=x,y,z) evolving under the 7 basic transformations (I_q, S_q, $2I_zS_z$, q=x,y,z) considered in this chapter.

3.5 Consider the coherence transfer operation: $2I_xS_z \xrightarrow{\beta(I_\phi + S_\phi)} -2I_zS_x$. Determine the flip angle (β) and phase (ϕ = +x,+y,−x or −y) of the pulse.

3.6 Assuming that spectra are phased such that I_y corresponds to absorption-mode lineshapes, sketch the spectra that correspond to the following two-spin product operators: (a) I_y, (b) I_x, (c) $2I_yS_z$, (d) $2I_xS_z$, (e) $\frac{1}{2}(I_y + 2I_yS_z)$, (f) $\frac{1}{2}(I_y - 2I_yS_z)$. Assume $J_{IS} > 0$.

3.7 Section 3.4 analyses the effect of a spin echo pulse sequence on a homonuclear two-spin system. Repeat this analysis using $\pi J_{IS}\tau = \pi/4$ from the outset.

3.8 Calculate the effect of the spin echo sequence $\tau - 180^\circ_y - \tau$ on $2I_xS_z$ in a homonuclear two-spin system. Hence determine the effect of the double spin echo sequence $90^\circ_x - \tau - 180^\circ_y - 2\tau - 180^\circ_y - \tau$ on the initial state I_z. Use previous results to simplify your calculations as much as possible.

Product operators II

4.1 Introduction

We now start to use product operators to describe pulse sequences that cannot be satisfactorily understood using the vector model. In this chapter we deal with one-dimensional experiments.

4.2 INEPT

INEPT: Insensitive Nuclei Enhanced by Polarization Transfer. NMR abounds with acronyms: not all are as ironic, whimsical, or even as tasteful as this one.

Recall that γ, the magnetogyric ratio of a nucleus, determines the energy-level splitting produced by a magnetic field and hence the population difference at equilibrium.

The INEPT experiment uses the magnetization of high-γ nuclei (such as ^{1}H, ^{19}F, ^{31}P, etc.) to enhance the weak NMR signals of low-γ nuclei (such as ^{13}C, ^{15}N, ^{57}Fe, etc.). The INEPT pulse sequence is also found as an element in many two-dimensional experiments such as the old-fashioned heteronuclear shift correlation and more modern 'inverse' techniques which have ^{13}C or ^{15}N magnetization evolving during t_1.

The initial (thermal equilibrium) state of the IS spin system can be written $aI_z + bS_z$, where the coefficients a and b are there to remind us that in a heteronuclear spin system the z-magnetizations associated with spins I and S have inherently different strengths. For example, for ^{1}H and ^{13}C, $a \approx 4b$ (because in this case $\gamma_\text{I} \approx 4\gamma_\text{S}$).

So far we have only shown that spin echoes refocus evolution under the resonance offset for particular initial states, but it can be shown that this refocusing occurs for *any* initial state, and so it is safe to neglect the resonance offset during calculations as done here. This point is explored in more detail in Part B.

The experiment (Fig. 4.1) starts with a 90°_x pulse on the I spin:

$$aI_z + bS_z \xrightarrow{90^\circ I_x} -aI_y + bS_z. \tag{4.1}$$

We now use the first of many short cuts that can be introduced to make product operator calculations less tedious. The INEPT sequence contains a spin echo: therefore, to calculate the state of the spin system at the end of the second τ period we can assume that there is no resonance offset Ω_I and just have the system evolving under the J-coupling, J_IS. The order in which the two 180° pulses are treated is irrelevant:

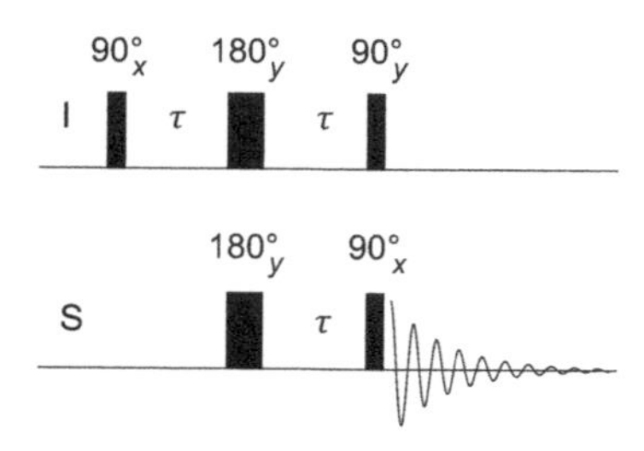

Fig. 4.1 The INEPT pulse sequence.

$$\begin{aligned}
&\xrightarrow{\pi J_\text{IS}\tau 2I_zS_z} -aI_y\cos(\pi J_\text{IS}\tau) + a2I_xS_z\sin(\pi J_\text{IS}\tau) + bS_z \\
&\xrightarrow{180^\circ I_y} -aI_y\cos(\pi J_\text{IS}\tau) - a2I_xS_z\sin(\pi J_\text{IS}\tau) + bS_z \\
&\xrightarrow{180^\circ S_y} -aI_y\cos(\pi J_\text{IS}\tau) + a2I_xS_z\sin(\pi J_\text{IS}\tau) - bS_z \\
&\xrightarrow{\pi J_\text{IS}\tau 2I_zS_z} -aI_y\cos^2(\pi J_\text{IS}\tau) + a2I_xS_z\cos(\pi J_\text{IS}\tau)\sin(\pi J_\text{IS}\tau) \\
&\qquad + a2I_xS_z\sin(\pi J_\text{IS}\tau)\cos(\pi J_\text{IS}\tau) + aI_y\sin^2(\pi J_\text{IS}\tau) - bS_z \\
&\qquad = -aI_y\cos(2\pi J_\text{IS}\tau) + a2I_xS_z\sin(2\pi J_\text{IS}\tau) - bS_z.
\end{aligned} \tag{4.2}$$

Note that apart from the factor of a the first two terms in Eqn 4.2 are identical to those in Eqn 3.19, and the result derived there could have been 'reused' here without explicitly repeating the calculation. This is a general feature of product operator calculations.

For the INEPT sequence the free precession interval τ is set as near as possible to $1/(4J_{\text{IS}})$, where J_{IS} is the coupling between the high-γ spin I and the low-γ spin S. For example, the one-bond ^{13}C–^{1}H J-coupling is typically about 150 Hz, giving $\tau = 1.67$ ms. Now if we substitute in the value $\tau = 1/(4J_{\text{IS}})$ we find that the purpose of the τ intervals is to allow the spin I doublet to evolve into a purely *antiphase* state, which the 90° pulses transfer from I to S:

$$= a2I_xS_z - bS_z \xrightarrow{90^\circ I_y} -a2I_zS_z - bS_z \xrightarrow{90^\circ S_x} a2I_zS_y + bS_y. \tag{4.3}$$

Both of these terms correspond to S-spin signals. The second represents in-phase y-magnetization and has the normal (low) intensity. The first is antiphase y-magnetization of spin S but its intensity factor is that of spin I. Thus, if I is ^{1}H and S is ^{13}C, the first term has approximately four times the intensity of the second, while if I is ^{1}H and S is ^{15}N, a is approximately ten times b. As shown in Fig. 4.2, we can think of each of these terms contributing towards the final S-spin spectrum.

To be more precise, as γ has opposite signs for ^{1}H and ^{15}N, the ^{15}N populations are *inverted* in comparison with ^{1}H. Thus $b \approx -0.1a$.

The asymmetry in the S spectrum can be removed by *phase cycling*, which is used to suppress the unenhanced spin S magnetization (the bS_y term). It is easily seen that if a second experiment is performed with the 90°_y pulse on the I spin replaced with a 90°_{-y} pulse then the final state of the system is $-a2I_zS_y + bS_y$. Thus, if this second spectrum is subtracted from the first, the final result is

$$(a2I_zS_y + bS_y) - (-a2I_zS_y + bS_y) = (2a)2I_zS_y. \tag{4.4}$$

This represents a purely antiphase S spectrum enhanced by a factor $a/b = \gamma_{\text{I}}/\gamma_{\text{S}}$. Nuclei with high γ usually relax faster than those with low γ and, since the

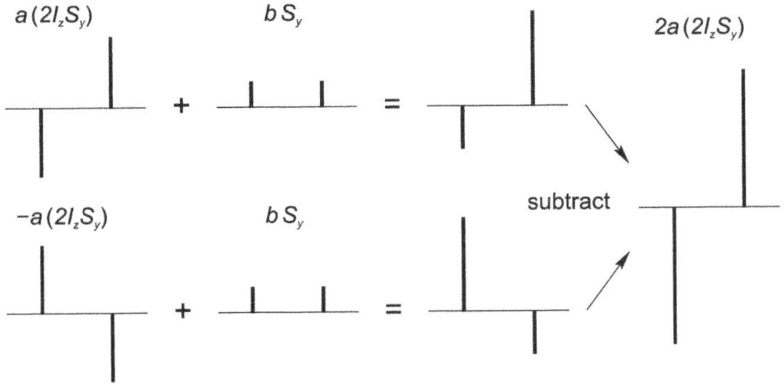

Fig. 4.2 In the INEPT experiment the bS_y contribution to the S-spin signal results in an asymmetric spectrum. Phase cycling can be used to remove the bS_y contribution if desired.

The additional factor of two enhancement in Eqn 4.4 is not interesting as it simply arises from doing the experiment twice.

repetition time of the INEPT sequence is determined by the relaxation of I magnetization, more free induction decays can be acquired and averaged in a given total experiment time; this represents an additional sensitivity advantage.

The antiphase signals observed in INEPT spectra do, however, have one significant disadvantage: heteronuclear decoupling, which would simplify the appearance of the spectrum, cannot straightforwardly be combined with INEPT. Removing the coupling during the free induction decay of the S spin causes the two lines, with equal and opposite intensities, to coincide and cancel completely. This problem can be overcome using more complicated sequences, as described in Section 4.5 and explored in the exercises at the end of this chapter.

4.3 **Multiple-quantum coherence**

The term 'coherence' is used to indicate the precessing states that involve at least one *x* or *y* operator, and includes I_x, I_y, S_x, and S_y, which represent single-quantum coherences. See also Chapters 6 and 8. Note that zero-quantum coherences are included in the category of multiple-quantum coherences.

The two-spin product operators, $2I_xS_x$, $2I_yS_x$, $2I_xS_y$, and $2I_yS_y$, which were neglected in Chapter 3, represent *multiple-quantum coherence*. Coherences are related to NMR transitions, and can be classified according to the value of $|\Delta m|$, the difference in magnetic quantum numbers of the two energy levels linked by the coherence. Coherences with $|\Delta m| \neq 1$ are not directly observable in an NMR experiment as they do not possess a magnetic dipole moment and so induce no current in the receiver coil. They cannot be described using the vector model.

From inspection of the energy-level diagram for a two-spin system (Fig. 4.3) we would expect four multiple-quantum coherences. The transitions DQ_x and DQ_y are the two double-quantum coherences (90° out-of-phase with respect to one another), while ZQ_x and ZQ_y are the two zero-quantum

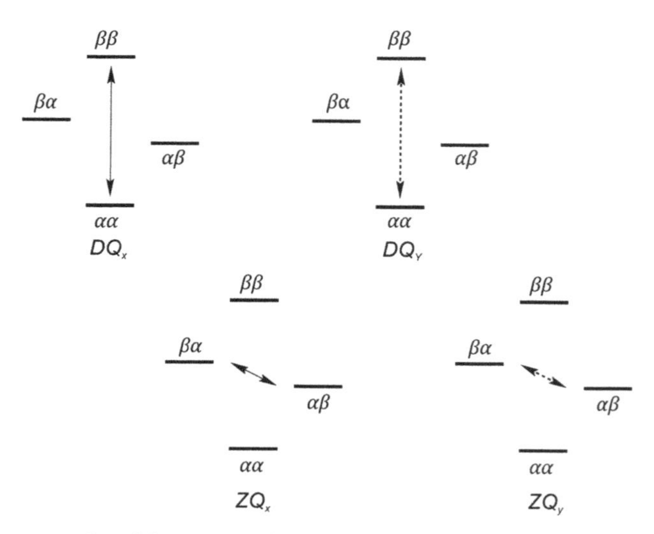

Fig. 4.3 Representation of the two-spin double- and zero-quantum coherence product operators in terms of the energy-level diagram. The dotted lines are a reminder of the 90° phase difference between *x* and *y* operators.

coherences. Unfortunately, all four are linear combinations of two-spin product operators:

$$DQ_x = \tfrac{1}{2}(2I_xS_x - 2I_yS_y) \qquad DQ_y = \tfrac{1}{2}(2I_yS_x + 2I_xS_y)$$
$$ZQ_x = \tfrac{1}{2}(2I_xS_x + 2I_yS_y) \qquad ZQ_y = \tfrac{1}{2}(2I_yS_x - 2I_xS_y). \tag{4.5}$$

Thus, for example, a two-spin operator $2I_xS_x$ corresponds to both double- and zero-quantum coherences: $2I_xS_x = DQ_x + ZQ_x$. Using Eqn 3.4 for both I and S, the evolution of these operators is fairly easily seen to be:

$$DQ_x \xrightarrow{\Omega_I t I_z + \Omega_S t S_z} DQ_x \cos[(\Omega_I + \Omega_S)t] + DQ_y \sin[(\Omega_I + \Omega_S)t]$$
$$ZQ_x \xrightarrow{\Omega_I t I_z + \Omega_S t S_z} ZQ_x \cos[(\Omega_I - \Omega_S)t] + ZQ_y \sin[(\Omega_I - \Omega_S)t] \tag{4.6}$$

and similarly for DQ_y and ZQ_y. Unsurprisingly, double-quantum coherences evolve at the sum of the two underlying frequencies, while zero-quantum coherences evolve at the difference frequency. Less obviously, IS zero- and double-quantum coherences do not evolve under the influence of the coupling J_{IS}. We will demonstrate this later (see Section 9.7).

Perhaps the simplest NMR experiment to make use of multiple-quantum coherence is a one-dimensional *double-quantum filter* (Fig. 4.4). The symbol ϕ outside the brackets around the first three pulses indicates that a phase-shift will be applied to these pulses on successive acquisitions as part of a phase cycle. Double-quantum coherence can be excited in multi-spin systems but not in isolated spin-$\tfrac{1}{2}$ nuclei. Thus a double-quantum filter can be used in ^{1}H spectroscopy to suppress solvent resonances (e.g. H_2O) and, under the acronym INADEQUATE, in ^{13}C spectroscopy, to suppress the signal from the 1% of molecules that contain a single ^{13}C nucleus in order to observe the 0.01% that contain two J-coupled ^{13}C spins.

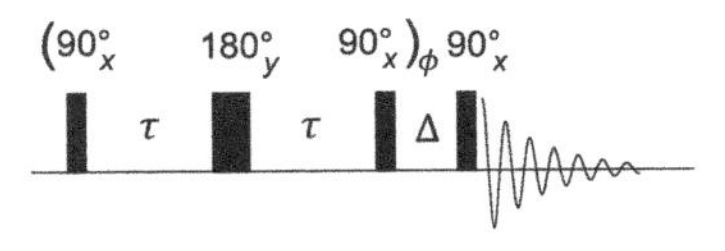

Fig. 4.4 Pulse sequence for the double-quantum filter (or INADEQUATE) experiment.

INADEQUATE: Incredible Natural Abundance Double Quantum Transfer Experiment. This experiment can be converted to a two-dimensional experiment by replacing the fixed delay Δ by a variable delay t_1.

If we imagine a three-spin system consisting of two J-coupled spins I and S and a third, isolated, spin R whose signal we wish to suppress, then at the end of the second τ delay the state of the system is given by:

$$-(I_y + S_y)\cos(2\pi J_{IS}\tau) + (2I_xS_z + 2I_zS_x)\sin(2\pi J_{IS}\tau) - R_y. \tag{4.7}$$

Apart from the last term, this is just Eqn 3.20.

As in INEPT, the delay τ is chosen to be $1/(4J_{IS})$ to maximize the antiphase component:

$$(2I_xS_z + 2I_zS_x) - R_y. \tag{4.8}$$

The second 90° pulse now creates a state of pure IS double-quantum coherence (DQ_y):

$$\xrightarrow{90^\circ(I_x + S_x + R_x)} -(2I_xS_y + 2I_yS_x) - R_z. \tag{4.9}$$

The interval Δ is a very short delay (typically a few microseconds) to allow the spectrometer to change the pulse phases; any evolution during Δ can safely be neglected. The final 90° pulse then reconverts the IS double-quantum coherence back into observable antiphase magnetization:

$$\xrightarrow{90^\circ(I_x+S_x+R_x)} -(2I_xS_z+2I_zS_x)+R_y. \tag{4.10}$$

But no discrimination between the coupled spins I and S and the isolated spin R has yet been achieved. We must perform a phase cycle that selects double-quantum coherence during the delay Δ and rejects single-quantum coherence (directly observable magnetization). This is done by successively adding $\phi = 90^\circ$ to the phase of the first three pulses over four experiments:

	Experiment	*Final state of system*
(1)	$90^\circ_{+x} - \tau - 180^\circ_{+y} - \tau - 90^\circ_{+x}\ 90^\circ_x$	$-(2I_xS_z+2I_zS_x)+R_y$
(2)	$90^\circ_{+y} - \tau - 180^\circ_{-x} - \tau - 90^\circ_{+y}\ 90^\circ_x$	$+(2I_xS_z+2I_zS_x)+R_y$
(3)	$90^\circ_{-x} - \tau - 180^\circ_{-y} - \tau - 90^\circ_{-x}\ 90^\circ_x$	$-(2I_xS_z+2I_zS_x)+R_y$
(4)	$90^\circ_{-y} - \tau - 180^\circ_{+x} - \tau - 90^\circ_{-y}\ 90^\circ_x$	$+(2I_xS_z+2I_zS_x)+R_y$. (4.11)

Therefore if, instead of merely adding the free induction decays produced by these four experiments, one takes the combination (1) – (2) + (3) – (4), the final result is:

$$(1)-(2)+(3)-(4) = -4(2I_xS_z+2I_zS_x). \tag{4.12}$$

In this way, the resonance of spin R does not appear in the final spectrum (Fig. 4.5). Although only two steps are necessary to cancel the R_y signal, all four are essential to avoid problems associated with the finite duration of the Δ delay. A much more general way of describing phase cycling will be presented in Chapter 6.

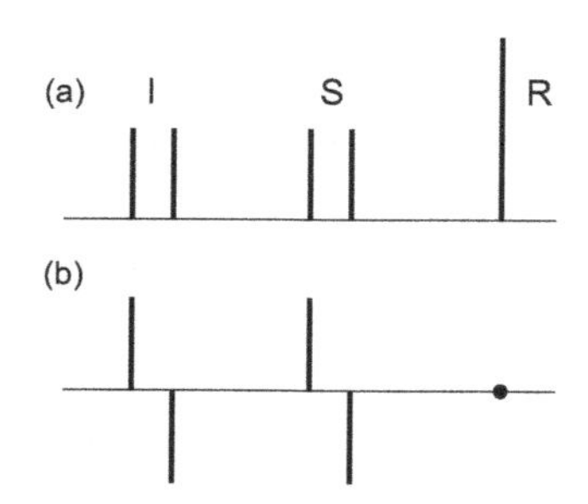

Fig. 4.5 Schematic spectra of an ISR spin system (a) following a 90° pulse and (b) after a double-quantum filter has been used to suppress the resonance of the uncoupled R spin.

4.4 Multi-spin systems

Systems of three *J*-coupled spin-$\frac{1}{2}$ nuclei, I, S, and R, are sufficiently complicated to show most of the various multiplet and flip-angle effects encountered in one- and two-dimensional NMR. There is rarely any need to do product operator calculations for four- and five-spin systems; the result can usually be predicted from knowledge of the three-spin case. There are 64 product operators for the ISR system and the reader should, if necessary, be able to write all of them down, deriving them from the two-spin operators in the same way as these were obtained from the one-spin operators.

We do perform one calculation on a four-spin system in Section 4.5.

The ISR spin system is described by three offset frequencies, Ω_I, Ω_S, and Ω_R, and three *J*-couplings, J_{IS}, J_{IR}, and J_{RS}. The new feature in the product operator description of an ISR system is the three-spin operators involving I, S, and R.

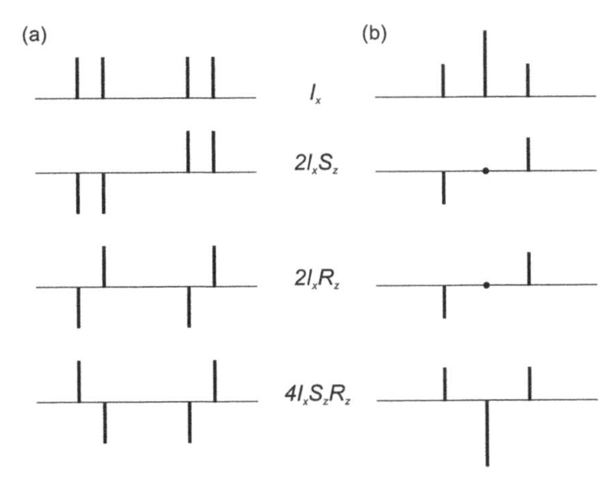

Fig. 4.6 Representation of the single-quantum I-spin product operators in an ISR spin system with (a) $J_{IS} > J_{IR}$ and (b) $J_{IS} = J_{IR}$.

These appear as a result of (say) I_x magnetization evolving under the influence of both J_{IS} and J_{IR}:

$$\begin{aligned} I_x &\xrightarrow{\pi J_{IS} t 2I_zS_z} I_x \cos(\pi J_{IS}t) + 2I_yS_z \sin(\pi J_{IS}t) \\ &\xrightarrow{\pi J_{IR} t 2I_zR_z} I_x \cos(\pi J_{IS}t)\cos(\pi J_{IR}t) + 2I_yR_z \cos(\pi J_{IS}t)\sin(\pi J_{IR}t) \\ &\quad + 2I_yS_z \sin(\pi J_{IS}t)\cos(\pi J_{IR}t) - 4I_xS_zR_z \sin(\pi J_{IS}t)\sin(\pi J_{IR}t). \end{aligned} \tag{4.13}$$

It is not always particularly useful to relate the three-spin product operators to the vector model or to energy-level diagrams, which are a little too complicated to be an intuitive aid. We should now be sufficiently familiar with the product operators to accept a description in words:

$4I_xS_zR_z$	x-magnetization of spin I, antiphase with respect to spins S and R
$4I_xS_xR_z$	mixture of IS double- and zero-quantum coherences, both antiphase with respect to spin R
$4I_xS_xR_x$	mixture of ISR triple-quantum coherence and unobservable ISR three-spin single-quantum coherences
$4I_zS_zR_z$	a highly ordered population state

It is useful, however, to relate the single-quantum product operators that occur in an ISR spin system to schematic spectra (Fig. 4.6). Any *J*-coupling between S and R is irrelevant to these I-spin multiplets.

4.5 **DEPT**

DEPT: Distortionless Enhancement by Polarization Transfer. It has now been largely replaced by the *multiplicity edited HSQC* experiment.

The DEPT experiment (Fig. 4.7) is a good example with which to end this chapter. Although it is no longer a popular or widely-used NMR experiment, its product

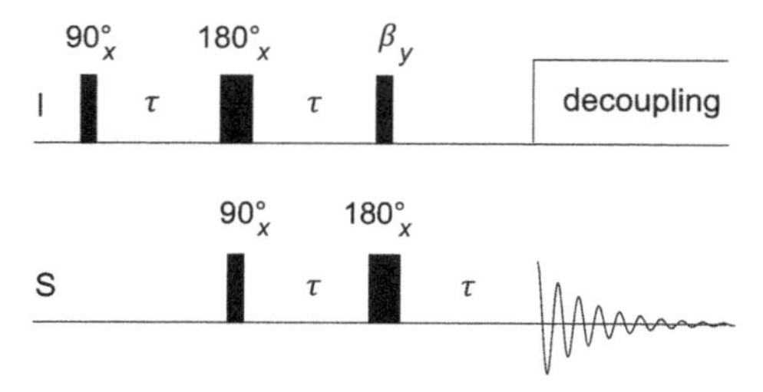

Fig. 4.7 The DEPT pulse sequence.

operator description, involving as it does multiple-quantum coherence in multi-spin systems, draws together many of the themes introduced above. DEPT can be used to 'edit' ^{13}C NMR spectra into sub-spectra containing only CH, CH_2, or CH_3 groups. Organic chemists find this sort of thing useful.

The various I spins are *magnetically equivalent*, and as a consequence appear not to be coupled to each other, as explained in Chapter 10. Here we simply neglect any couplings between the I spins.

We can start by following the magnetization of just one proton (I_1), calling the others (if present) I_2 and I_3, all with the same offset Ω_I and coupled to a ^{13}C nucleus S with the same coupling constant J_{IS}:

This is just Eqn 3.12 in a slightly different notation.

$$I_{1z}\xrightarrow{90°I_x}\xrightarrow{\Omega_I\tau I_z}\xrightarrow{\pi J_{IS}\tau 2I_{1z}S_z}$$
$$-I_{1y}\cos\Omega_I\tau\cos(\pi J_{IS}\tau)+2I_{1x}S_z\cos\Omega_I\tau\sin(\pi J_{IS}\tau)$$
$$+I_{1x}\sin\Omega_I\tau\cos(\pi J_{IS}\tau)+2I_{1y}S_z\sin\Omega_I\tau\sin(\pi J_{IS}\tau). \quad (4.14)$$

Equivalent terms can, of course, arise from I_{2z} and I_{3z}, if present, but these will behave in exactly the same way.

The τ period is set to $1/(2J_{IS})$ so that we get:

$$2I_{1x}S_z\cos\Omega_I\tau+2I_{1y}S_z\sin\Omega_I\tau. \quad (4.15)$$

The next two pulses create a state of heteronuclear multiple-quantum (ZQ and DQ) coherence:

$$\xrightarrow{180°I_x}2I_{1x}S_z\cos\Omega_I\tau-2I_{1y}S_z\sin\Omega_I\tau$$
$$\xrightarrow{90°S_x}-2I_{1x}S_y\cos\Omega_I\tau+2I_{1y}S_y\sin\Omega_I\tau. \quad (4.16)$$

This multiple-quantum coherence evolves during the next free precession interval under the influence of both I and S chemical shifts:

$$\xrightarrow{\Omega_I\tau I_z}-2I_{1x}S_y\cos\Omega_I\tau\cos\Omega_I\tau-2I_{1y}S_y\cos\Omega_I\tau\sin\Omega_I\tau$$
$$+2I_{1y}S_y\sin\Omega_I\tau\cos\Omega_I\tau-2I_{1x}S_y\sin\Omega_I\tau\sin\Omega_I\tau$$
$$=-2I_{1x}S_y$$
$$\xrightarrow{\Omega_S\tau S_z}-2I_{1x}S_y\cos\Omega_S\tau+2I_{1x}S_x\sin\Omega_S\tau. \quad (4.17)$$

The I_1-spin chemical shift has been refocused, but not—yet—the S-spin shift. These coherences, which involve coherent terms on spin S, will also evolve under the influence of the $2I_{2z}S_z$ and $2I_{3z}S_z$ couplings, if present. As noted in Section 4.3, the J-coupling between I_1 and S does not cause the multiple-quantum coherences between the I_1 and S spins to evolve. So the result of the evolution under J-coupling during the second τ period depends on whether we are describing a CH, CH_2, or CH_3 group:

$$\begin{aligned}
&\text{CH} \xrightarrow{\qquad\qquad\qquad} -2I_{1x}S_y\cos\Omega_S\tau + 2I_{1x}S_x\sin\Omega_S\tau,\\
&\text{CH}_2 \xrightarrow{\pi J_{IS}\tau 2I_{2z}S_z\ (\tau=1/2J_{IS})} 4I_{1x}I_{2z}S_x\cos\Omega_S\tau + 4I_{1x}I_{2z}S_y\sin\Omega_S\tau,\\
&\text{CH}_3 \xrightarrow{\pi J_{IS}\tau 2I_{2z}S_z\ (\tau=1/2J_{IS})} 4I_{1x}I_{2z}S_x\cos\Omega_S\tau + 4I_{1x}I_{2z}S_y\sin\Omega_S\tau\\
&\qquad \xrightarrow{\pi J_{IS}\tau 2I_{3z}S_z\ (\tau=1/2J_{IS})} 8I_{1x}I_{2z}I_{3z}S_y\cos\Omega_S\tau - 8I_{1x}I_{2z}I_{3z}S_x\sin\Omega_S\tau.
\end{aligned} \tag{4.18}$$

Remember that I_1, I_2, and I_3 are *equivalent*, so that all three couplings have the same value, J_{IS}.

The next two pulses reconvert these heteronuclear multiple-quantum coherences into observable single-quantum coherences (non-observable terms are not shown below):

$$\begin{aligned}
&\text{CH} \xrightarrow{\beta I_y} \xrightarrow{180°S_x} -2I_{1z}S_y\cos\Omega_S\tau\sin\beta\\
&\qquad\qquad - 2I_{1z}S_x\sin\Omega_S\tau\sin\beta,\\
&\text{CH}_2 \xrightarrow{\beta I_y} \xrightarrow{180°S_x} -4I_{1z}I_{2z}S_x\cos\Omega_S\tau\sin\beta\cos\beta\\
&\qquad\qquad + 4I_{1z}I_{2z}S_y\sin\Omega_S\tau\sin\beta\cos\beta,\\
&\text{CH}_3 \xrightarrow{\beta I_y} \xrightarrow{180°S_x} 8I_{1z}I_{2z}I_{3z}S_y\cos\Omega_S\tau\sin\beta\cos^2\beta\\
&\qquad\qquad + 8I_{1z}I_{2z}I_{3z}S_x\sin\Omega_S\tau\sin\beta\cos^2\beta.
\end{aligned} \tag{4.19}$$

The final free-precession period leads to refocusing of the S-spin chemical shifts and converts the antiphase operators to in-phase states which will be observable in the presence of I-spin decoupling:

$$\begin{aligned}
&\text{CH} \xrightarrow{\Omega_S\tau S_z} \xrightarrow{\pi J_{IS}\tau\, 2I_{1z}S_z\,(\tau=1/2J_{IS})} S_x\sin\beta,\\
&\text{CH}_2 \xrightarrow{\Omega_S\tau S_z} \xrightarrow{\pi J_{IS}\tau\, 2I_{1z}S_z\,(\tau=1/2J_{IS})}\\
&\qquad \xrightarrow{\pi J_{IS}\tau\, 2I_{2z}S_z\,(\tau=1/2J_{IS})} S_x\sin\beta\cos\beta,\\
&\text{CH}_3 \xrightarrow{\Omega_S\tau S_z} \xrightarrow{\pi J_{IS}\tau\, 2I_{1z}S_z\,(\tau=1/2J_{IS})}\\
&\qquad \xrightarrow{\pi J_{IS}\tau\, 2I_{2z}S_z\,(\tau=1/2J_{IS})}\\
&\qquad \xrightarrow{\pi J_{IS}\tau\, 2I_{3z}S_z\,(\tau=1/2J_{IS})} S_x\sin\beta\cos^2\beta.
\end{aligned} \tag{4.20}$$

Thus, the final ^{13}C spectrum will, in general, contain the resonances of all three types of carbon atoms and, as with INEPT, there will be a signal enhancement proportional to the ratio γ_I/γ_S. However, the different flip angle dependencies allow one to edit the spectrum according to multiplicity. If three spectra A, B, and C are recorded with flip angles of $\beta = 45°$, 90°, and 135°, respectively, then spectrum B contains only CH signals, the difference (A–C) has only CH_2 signals, while a spectrum containing only CH_3 signals corresponds to the linear combination $[\sqrt{2}(A+C)-2B]$.

A second advantage of DEPT over INEPT is that the observable signals are in-phase rather than antiphase (hence the *distortionless* in the acronym). This is not just a matter of taste: restoring in-phase signals allows heteronuclear decoupling to be used with DEPT, simplifying the spectrum and concentrating all the signal intensity in a single line. Note that in a DEPT spectrum there is no reason to retain the coupling in order to distinguish between CH, CH_2, and CH_3 groups.

4.6 Summary

- Product operators allow experiments more complex than spin echoes to be analysed.
- Phase cycling can be used to simplify spectra by cancelling unwanted terms.
- Multiple-quantum coherences cannot be directly observed but can be made from, and turned back into, observable single-quantum coherences.
- Phase cycling can also be used to select particular multiple-quantum coherences.
- Polarization transfer (e.g. in INEPT and DEPT) from high-γ spins can be used to enhance the sensitivity of low-γ spins.
- Multi-spin systems can be treated by simple extension of two-spin product operator calculations.
- DEPT combines all these ideas to produce intensity-enhanced 'edited' spectra.

4.7 Exercises

Worked solutions to the exercises are available on the Online Resource Centre at www.oxfordtextbooks.co.uk/orc/hore2e/

4.1. Consider the INEPT pulse sequence in Fig. 4.1 applied to a two-spin system (I and S). Show that insertion of the pulse sequence element $\tau - 180°I_y 180°S_y - \tau$ immediately before the free induction decay, where τ has the same value as in the first part of the sequence, results in an enhanced *in-phase* spectrum of spin S.

4.2. Calculate the product operators arising from an INEPT sequence applied to a CH_2 group and draw the resulting enhanced spectrum. Would phase cycling work in the same way here as it does for a CH group?

4.3. Repeat Exercise 4.2 for a CH_3 group.

4.4. Would the INEPT pulse sequence in Fig. 4.1 still work if: (a) both 180°_y pulses were replaced by 180°_x pulses? (b) the final 90° S_x pulse were replaced by a 90° S_y pulse? (c) the final 90° I_y pulse were replaced by a 90° I_x pulse?

4.5. Prove the results given in Eqn 4.6.

4.6. What detectable signal results from $2I_zS_z \xrightarrow{\beta(I_x+S_x)}$? Sketch the form of the spectrum. What flip angle gives the maximum detectable signal?

4.7. Four new one-spin product operators can be defined by $I_\pm = I_x \pm iI_y$ and $S_\pm = S_x \pm iS_y$. Write $\frac{1}{2}(2I_xS_x \pm 2I_yS_y)$ in terms of $I_\pm$ and $S_\pm$ and interpret the results.

4.8. Show that the rules for combining DEPT spectra stated after Eqn 4.20 are correct.

4.9. What would happen to a quaternary carbon in a DEPT experiment?

5 Two-dimensional NMR

5.1 Introduction

The workings of a few of the more commonly used two-dimensional NMR experiments will now be described using product operators. Our aim will be to see how they work and to explain their chief features with respect to phase and multiplet structure.

5.2 COSY

COSY: Correlation Spectroscopy.

COSY (Fig. 5.1) is the archetypal two-dimensional NMR experiment. It was the first to be proposed (in 1971) and is still one of the most popular. It is used to correlate J-coupled spins for purposes of spectral assignment. We will see that a two-dimensional COSY spectrum contains two types of multiplets: *diagonal peaks* centred on the same chemical shift in the two frequency dimensions F_1 and F_2 and *cross peaks* centred on different chemical shifts in the two dimensions; the latter will only occur if the two chemical shifts correspond to a J-coupled pair of spins.

First consider a two-spin IS system. We will only calculate the evolution for spin I; S behaves identically, but with the labels I and S interchanged. Immediately after the second pulse we have:

$$\begin{aligned} I_z &\xrightarrow{90^\circ(I_x+S_x)} \xrightarrow{\Omega_I t_1 I_z + \Omega_S t_1 S_z} \xrightarrow{\pi J_{IS} t_1 2I_z S_z} \\ &\quad -I_y \cos\Omega_I t_1 \cos(\pi J_{IS} t_1) + 2I_x S_z \cos\Omega_I t_1 \sin(\pi J_{IS} t_1) \\ &\quad + I_x \sin\Omega_I t_1 \cos(\pi J_{IS} t_1) + 2I_y S_z \sin\Omega_I t_1 \sin(\pi J_{IS} t_1) \\ &\xrightarrow{90^\circ(I_x+S_x)} -I_z \cos\Omega_I t_1 \cos(\pi J_{IS} t_1) - 2I_x S_y \cos\Omega_I t_1 \sin(\pi J_{IS} t_1) \\ &\quad + I_x \sin\Omega_I t_1 \cos(\pi J_{IS} t_1) - 2I_z S_y \sin\Omega_I t_1 \sin(\pi J_{IS} t_1). \end{aligned} \tag{5.1}$$

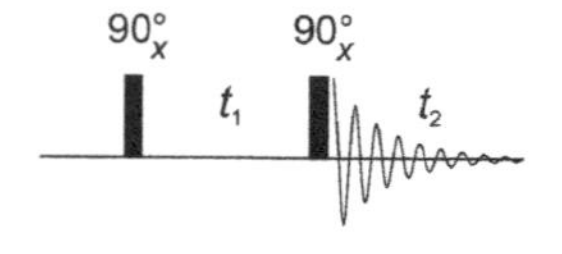

Fig. 5.1 The COSY pulse sequence.

Of the four terms, the first two (I_z and $2I_xS_y$) are not observable. The third (I_x) represents spin I magnetization that has evolved at the offset frequency Ω_I in t_1 and will evolve at the same frequency in t_2. It therefore gives rise to a diagonal peak. The final term ($2I_zS_y$), on the other hand, represents S magnetization that has evolved at the offset frequency Ω_I in t_1 but will evolve at offset Ω_S in t_2. It therefore corresponds to a COSY cross peak between spins I and S. Its dependence on $\sin\pi J_{IS}t_1$ shows that it will only be present if J_{IS} is non-zero.

We can see from the form of the operators I_x and $2I_zS_y$ that the diagonal peak will have an in-phase multiplet structure in F_2 while the cross peak will be antiphase with respect to the J_{IS} coupling. Also note that the two peaks will be 90° out-of-phase with respect to each other in F_2. Hence, if the cross peak is phased to be absorptive in this dimension, the diagonal peak will be dispersive. A schematic cross section parallel to F_2 (Fig. 5.2) perhaps shows this more clearly.

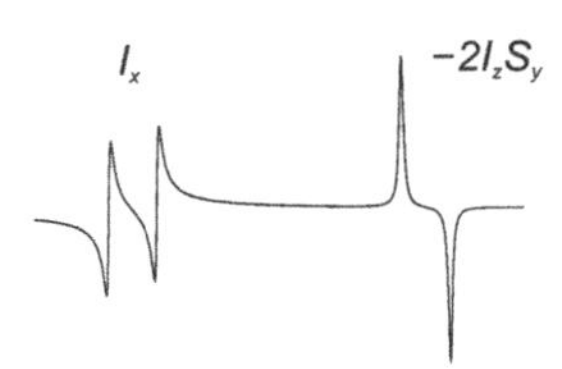

Fig. 5.2 Schematic cross section, parallel to the F_2 axis, through a COSY spectrum of an IS spin system.

But inspection of the operator terms that exist at the start of acquisition in t_2 only gives us the form of the spectrum in the F_2 dimension. To determine the appearance of the COSY spectrum in the F_1 dimension we must examine the modulation that these observables have acquired during t_1. If we expand the amplitudes of I_x and $2I_zS_y$ in Eqn 5.1 using trigonometric relations we find:

$$\begin{aligned} &I_x \sin\Omega_I t_1 \cos(\pi J_{IS}t_1) - 2I_zS_y \sin\Omega_I t_1 \sin(\pi J_{IS}t_1) \\ &\quad = I_x \tfrac{1}{2}\left[\sin([\Omega_I + \pi J_{IS}]t_1) + \sin([\Omega_I - \pi J_{IS}]t_1)\right] \\ &\quad - 2I_zS_y \tfrac{1}{2}\left[\cos([\Omega_I - \pi J_{IS}]t_1) - \cos([\Omega_I + \pi J_{IS}]t_1)\right]. \end{aligned} \tag{5.2}$$

$\sin(A \pm B) = \sin A \cos B \pm \cos A \sin B$
$\cos(A \pm B) = \cos A \cos B \mp \sin A \sin B$

The two frequencies $\Omega_I + \pi J_{IS}$ and $\Omega_I - \pi J_{IS}$ correspond to the two components of the I-spin doublet. In the I_x term the sines of these two frequencies have the same sign and so represent an *in-phase* doublet while in the $2I_zS_y$ term the corresponding cosines have opposite signs—an antiphase doublet. The fact that the respective modulations are sine and cosine indicates that in F_1 as well as F_2 there will be a 90° phase difference between the diagonal and cross peaks. Thus the multiplet structures and relative phases of the diagonal and cross peaks will be the same in both the F_1 and F_2 dimensions of a COSY spectrum. The appearance of the COSY spectrum of an IS spin system is therefore as shown schematically in Fig. 5.3. Finally note that both diagonal and cross peaks are *amplitude*-modulated, which means that the procedure outlined in Section 2.4, or something equivalent, will be needed to determine the sense of precession during t_1.

A disadvantage of *antiphase* cross peaks is that the positive and negative portions tend to overlap and cancel one another. This can be troublesome for large molecules which have broad lines; in such cases TOCSY (Section 5.5) provides a useful alternative.

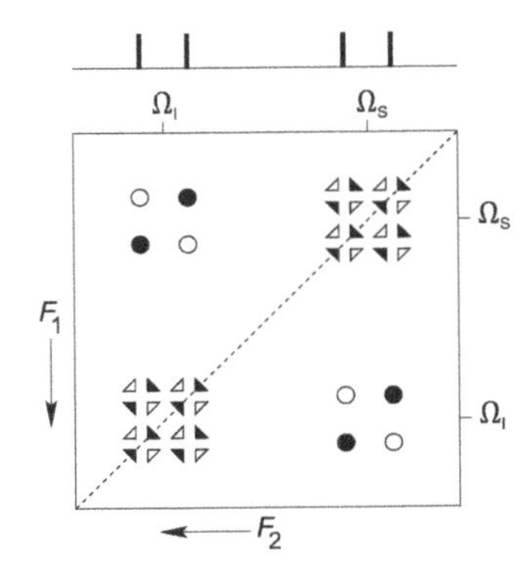

Fig. 5.3 Schematic COSY spectrum of an IS spin system, showing dispersive, in-phase diagonal peaks and absorptive, antiphase cross peaks. Solid and open symbols represent positive and negative peaks, respectively. Circles represent the double absorption lineshape (Fig. 2.6(a)); the double dispersion lineshape (Fig. 2.6(b)) is indicated by four triangles.

This analysis is easily extended to a three-spin ISR system. Starting with the I_z magnetization as before, ignoring evolution operators which have no effect on the state of the spin system, and omitting all of the intermediate steps:

$$\begin{aligned} &I_z \xrightarrow{90^\circ I_x} \xrightarrow{\Omega_I t_1 I_z} \xrightarrow{\pi J_{IS} t_1 2I_zS_z} \xrightarrow{\pi J_{IR} t_1 2I_zR_z} \xrightarrow{90^\circ(I_x+S_x+R_x)} \\ &- I_z \cos\Omega_I t_1 \cos(\pi J_{IS}t_1)\cos(\pi J_{IR}t_1) \\ &- 2I_xS_y \cos\Omega_I t_1 \sin(\pi J_{IS}t_1)\cos(\pi J_{IR}t_1) \\ &+ I_x \sin\Omega_I t_1 \cos(\pi J_{IS}t_1)\cos(\pi J_{IR}t_1) \qquad (1) \\ &- 2I_zS_y \sin\Omega_I t_1 \sin(\pi J_{IS}t_1)\cos(\pi J_{IR}t_1) \qquad (2) \\ &- 2I_xR_y \cos\Omega_I t_1 \cos(\pi J_{IS}t_1)\sin(\pi J_{IR}t_1) \\ &+ 4I_zS_yR_y \cos\Omega_I t_1 \sin(\pi J_{IS}t_1)\sin(\pi J_{IR}t_1) \\ &- 2I_zR_y \sin\Omega_I t_1 \cos(\pi J_{IS}t_1)\sin(\pi J_{IR}t_1) \qquad (3) \\ &- 4I_xS_yR_y \sin\Omega_I t_1 \sin(\pi J_{IS}t_1)\sin(\pi J_{IR}t_1). \end{aligned} \tag{5.3}$$

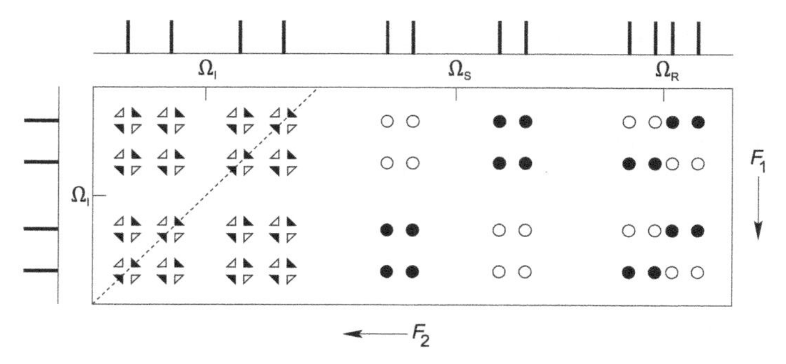

Fig. 5.4 Part of a schematic COSY spectrum of an ISR spin system, showing the diagonal and cross peaks centred at frequency Ω_I in the F_1 dimension. The dashed line indicates the part of the diagonal that passes through this section of the spectrum.

A modulation term such as $\sin\Omega_I t_1 \cos(\pi J_{IS} t_1)\cos(\pi J_{IR} t_1)$ can readily be seen to correspond to overall sine modulation as it is the product of one *odd* (sine) and two *even* (cosine) functions, making it an odd function overall. Hence it must expand (using the sine and cosine addition formulae given above) as a linear combination of four sine functions.

Of the eight terms, only those labelled (1), (2), and (3) give rise to observable signals. (1) is a diagonal peak. In the F_2 dimension it is in-phase with respect to both J_{IS} and J_{IR} and dispersive (both of these because the operator is I_x), while in the F_1 dimension it is also in-phase with respect to both J_{IS} and J_{IR} (cosine modulated by both couplings) and dispersive (overall sine modulated). (2) is an IS cross peak. In the F_2 dimension it is antiphase with respect to J_{IS} but in-phase with respect to J_{SR} and absorptive (both of these because the operator is $2I_zS_y$), while in the F_1 dimension it is also antiphase with respect to J_{IS} (sine modulated by J_{IS}) but in-phase with respect to J_{IR} (cosine modulated by J_{IR}) and absorptive (overall cosine modulated). Note that the antiphase structure arises from the *active* coupling (J_{IS} for the IS cross peak) while the passive couplings (J_{IR} and J_{SR}) give in-phase structure, and that the passive structure is different in the F_1 and F_2 dimensions of the cross peak. (3) is the analogous IR cross peak. Fig. 5.4 shows these three terms in the form of a (partial) schematic COSY spectrum.

5.3 DQF-COSY

DQF-COSY: Double-Quantum Filtered COSY.

A major drawback of COSY is that the diagonal peaks are rather intense because they are in-phase and cover a large area of the two-dimensional NMR spectrum on account of their dispersive nature. Thus cross peaks close to the diagonal can be obscured. A popular variant, DQF-COSY (Fig. 5.5), attempts to solve this problem by causing the diagonal peaks to be antiphase and *nearly* pure absorption. Any singlet resonances from isolated spins are also suppressed. As a result, DQF-COSY is often considered superior to the basic COSY experiment.

The phase ϕ of the first two pulses is cycled, as in the one-dimensional double-quantum filter (Section 4.3), to select only pure double-quantum coherence during the negligible interval Δ. For the I-spin magnetization of an IS spin system we find:

$$I_z \xrightarrow{90°I_x} \xrightarrow{\Omega_I t_1 I_z} \xrightarrow{\pi J_{IS} t_1 2I_zS_z} \xrightarrow{90°(I_x+S_x)} -2I_xS_y \cos\Omega_I t_1 \sin(\pi J_{IS} t_1) \quad (5.4)$$

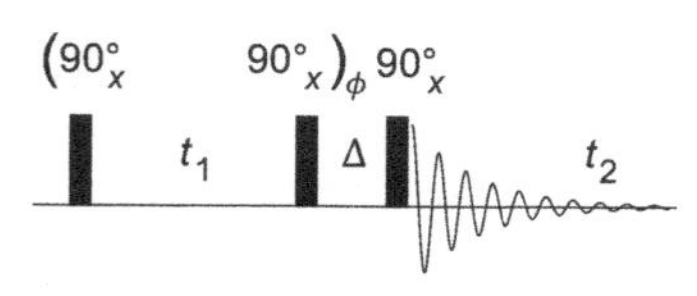

Fig. 5.5 The double-quantum filtered COSY pulse sequence.

where we have retained only the term containing the desired IS double-quantum coherence to be selected by the phase cycle. From Eqn 4.5, the product operator $2I_xS_y$ is equal to $DQ_y - ZQ_y$, the zero-quantum component of which will be removed by the phase cycle, so we can happily replace the $2I_xS_y$ by the operator $DQ_y = \frac{1}{2}(2I_xS_y + 2I_yS_x)$, to obtain:

$$\xrightarrow{\text{select DQ}} -\tfrac{1}{2}(2I_xS_y + 2I_yS_x)\cos\Omega_I t_1 \sin(\pi J_{IS} t_1)$$
$$\xrightarrow{90^\circ(I_x+S_x)} -\tfrac{1}{2}(2I_xS_z + 2I_zS_x)\cos\Omega_I t_1 \sin(\pi J_{IS} t_1). \qquad (5.5)$$

Both parts of this final state are observable. The first represents a diagonal peak (with an offset frequency of Ω_I in both t_1 and t_2) that will be antiphase with respect to J_{IS} in F_2 (because the operator is $2I_xS_z$) *and* in F_1 (sine modulation by J_{IS}). The second represents the IS cross peak (with an offset frequency of Ω_I in F_1 and Ω_S in F_2) which will also be antiphase with respect to J_{IS} in both F_1 and F_2. Finally, note that both diagonal and cross peaks have the same (amplitude-) modulation in t_1 and the same phase in t_2 and can therefore both be phased to pure absorption in both dimensions. The DQF-COSY spectrum of an IS spin system is therefore as shown in Fig. 5.6. The only disadvantage of DQF-COSY with respect to COSY is revealed by the factor of one half in Eqn 5.5; the sensitivity of the double-quantum filtered experiment is half that of the basic experiment.

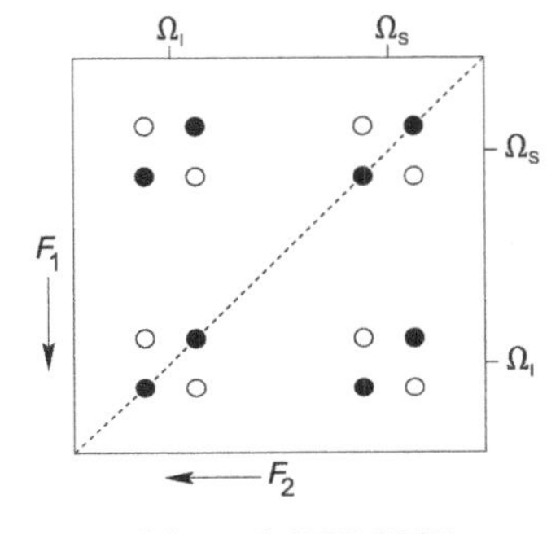

Fig. 5.6 Schematic DQF-COSY spectrum of an IS spin system, showing the absorptive, antiphase diagonal *and* cross peaks.

Another feature of DQF-COSY emerges when the product operator calculation is performed for an ISR spin system. After the second 90° pulse we find:

$$I_z \xrightarrow{90^\circ I_x} \xrightarrow{\Omega_I t_1 I_z} \xrightarrow{\pi J_{IS} t_1 2I_zS_z} \xrightarrow{\pi J_{IR} t_1 2I_zR_z} \xrightarrow{90^\circ(I_x+S_x+R_x)}$$
$$-2I_xS_y \cos\Omega_I t_1 \sin(\pi J_{IS} t_1)\cos(\pi J_{IR} t_1)$$
$$-2I_xR_y \cos\Omega_I t_1 \cos(\pi J_{IS} t_1)\sin(\pi J_{IR} t_1)$$
$$+4I_zS_yR_y \cos\Omega_I t_1 \sin(\pi J_{IS} t_1)\sin(\pi J_{IR} t_1) \qquad (5.6)$$

where again we have only retained the terms containing double-quantum coherence. As for the IS spin system, because of the use of a phase cycle we can now replace these mixed operators with their pure double-quantum equivalents:

$$\xrightarrow{\text{select DQ}} -\tfrac{1}{2}(2I_xS_y + 2I_yS_x)\cos\Omega_I t_1 \sin(\pi J_{IS} t_1)\cos(\pi J_{IR} t_1)$$
$$-\tfrac{1}{2}(2I_xR_y + 2I_yR_x)\cos\Omega_I t_1 \cos(\pi J_{IS} t_1)\sin(\pi J_{IR} t_1)$$
$$-\tfrac{1}{2}(4I_zS_xR_x - 4I_zS_yR_y)\cos\Omega_I t_1 \sin(\pi J_{IS} t_1)\sin(\pi J_{IR} t_1)$$
$$\xrightarrow{90^\circ(I_x+S_x+R_x)} -\tfrac{1}{2}(2I_xS_z + 2I_zS_x)\cos\Omega_I t_1 \sin(\pi J_{IS} t_1)\cos(\pi J_{IR} t_1)$$
$$-\tfrac{1}{2}(2I_xR_z + 2I_zR_x)\cos\Omega_I t_1 \cos(\pi J_{IS} t_1)\sin(\pi J_{IR} t_1)$$
$$+\tfrac{1}{2}(4I_yS_xR_x - 4I_yS_zR_z)\cos\Omega_I t_1 \sin(\pi J_{IS} t_1)\sin(\pi J_{IR} t_1). \qquad (5.7)$$

The first pair of product operators gives rise to an absorptive, antiphase (with respect to J_{IS}) I-spin diagonal peak and an absorptive, antiphase IS cross peak. The second pair will give rise to another absorptive, antiphase (but this time with respect to J_{IR}) contribution to the I-spin diagonal peak and an absorptive, antiphase IR cross peak. All this could have been extrapolated from the result for the IS spin system, but a novel feature is represented by the final pair of three-spin product operators. The first term of the pair is not observable but the second, $4I_yS_zR_z$, represents another contribution to the I-spin diagonal peak. It is doubly antiphase, however—antiphase with respect to both J_{IS} and J_{IR}—in both dimensions and is also *dispersive* (relative to the other peaks) in both dimensions. Thus, in systems of three or more spins, there *is* a dispersive contribution to the diagonal of a DQF-COSY. However, because it is antiphase with respect to two couplings, the cancellation of positive and negative intensities means that it is normally weaker than the absorptive components, and the diagonal peak multiplets remain largely absorptive.

5.4 NOESY

NOESY: Nuclear Overhauser Effect (or Enhancement) Spectroscopy.

NOESY (Fig. 5.7) is one of the oldest two-dimensional NMR experiments and still one of the most widely used. In contrast to COSY, which correlates J-coupled spins, the existence of a NOESY cross peak indicates a *nuclear Overhauser effect* between the two nuclei, implying that they can be no more than a few hundred picometres apart.

The nuclear Overhauser effect (NOE) is the name given to the transfer of z-magnetization from one spin to another by cross relaxation induced by the dipolar interaction of the two spins. It gives information on the 'through-space' separation of the nuclei; see Hore (2015).

Consider a two-spin IS system. Starting with z-magnetization of spin I, the effect of the first part of the pulse sequence is identical to COSY (Eqn 5.1):

$$\begin{aligned} I_z &\xrightarrow{90^\circ I_x} \xrightarrow{\Omega_I t_1 I_z} \xrightarrow{\pi J_{IS} t_1 2I_zS_z} \xrightarrow{90^\circ(I_x+S_x)} \\ &-I_z\cos\Omega_I t_1\cos(\pi J_{IS}t_1) - 2I_xS_y\cos\Omega_I t_1\sin(\pi J_{IS}t_1) \\ &+I_x\sin\Omega_I t_1\cos(\pi J_{IS}t_1) - 2I_zS_y\sin\Omega_I t_1\sin(\pi J_{IS}t_1). \end{aligned} \tag{5.8}$$

The phase cycling is designed to select only coherence order $p=0$ during τ_m, i.e. populations and zero-quantum coherences:

$$\xrightarrow{\text{select } p=0} -I_z\cos\Omega_I t_1\cos(\pi J_{IS}t_1) + ZQ_y\cos\Omega_I t_1\sin(\pi J_{IS}t_1). \tag{5.9}$$

The full meaning of the terms coherence and coherence order are explained in Chapters 4 and 6.

During the mixing period τ_m, z-magnetization is passed from I to S either by cross relaxation (i.e. NOE) or chemical exchange. In general terms, the mixing period results in the transformation

$$I_z \longrightarrow aI_z + bS_z \tag{5.10}$$

where the coefficients a and b are numbers that depend on τ_m and the efficiency of the polarization transfer. Appendix A gives some insight into the origin of this effect. Remembering that the zero-quantum coherence in Eqn 5.9 precesses during τ_m, we find:

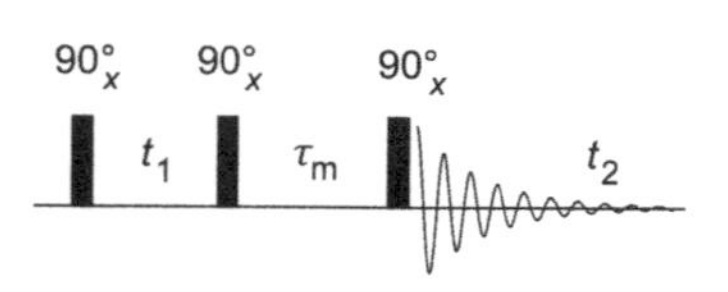

Fig. 5.7 The NOESY pulse sequence. In this, and all subsequent pulse sequences, the use of phase cycling is *not* indicated (cf. Fig. 5.5).

$$\begin{aligned} &\xrightarrow{(\Omega_I I_z+\Omega_S S_z)\tau_m} -aI_z\cos\Omega_I t_1\cos(\pi J_{IS}t_1) - bS_z\cos\Omega_I t_1\cos(\pi J_{IS}t_1) \\ &\quad + ZQ_y\cos\Omega_I t_1\sin(\pi J_{IS}t_1)\cos([\Omega_I - \Omega_S]\tau_m) \\ &\quad - ZQ_x\cos\Omega_I t_1\sin(\pi J_{IS}t_1)\sin([\Omega_I - \Omega_S]\tau_m), \end{aligned} \tag{5.11}$$

and after the final pulse:

$$\xrightarrow{90^\circ(I_x+S_x)} aI_y\cos\Omega_I t_1\cos(\pi J_{IS}t_1)+bS_y\cos\Omega_I t_1\cos(\pi J_{IS}t_1)$$
$$+\tfrac{1}{2}(2I_zS_x-2I_xS_z)\cos\Omega_I t_1\sin(\pi J_{IS}t_1)\cos([\Omega_I-\Omega_S]\tau_m)$$
$$-\tfrac{1}{2}(2I_xS_x-2I_zS_z)\cos\Omega_I t_1\sin(\pi J_{IS}t_1)\sin([\Omega_I-\Omega_S]\tau_m). \quad (5.12)$$

The eagle-eyed will have spotted that NOESY (Fig. 5.7) and DQF-COSY (Fig. 5.5) have seemingly identical pulse sequences. They give different spectra because of their different phase cycles which select z-magnetization and double-quantum coherence respectively during the second delay (τ_m or Δ). See Chapter 6 for more details.

The first term, aI_y, is an in-phase diagonal peak which is absorptive in both dimensions. The second, bS_y, is an in-phase IS cross peak, also absorptive in both dimensions. The third, $2I_zS_x-2I_xS_z$, represents unwanted contributions to the diagonal and cross peaks originating from the zero-quantum terms in Eqn 5.11 and these are antiphase and dispersive in both F_1 and F_2 dimensions. These peaks can be suppressed by recording and co-adding several NOESY spectra with a range of values of the mixing time τ_m or by more sophisticated means. Alternatively, in the spectra of large molecules, they can often be ignored because the zero-quantum relaxation rate is very fast and the antiphase contributions tend to cancel. In any case, they do not affect the integrated volume of the cross peaks. The fourth term in Eqn 5.12 is not observable.

NOE cross peaks *only* appear if there is an exchange of z-magnetization during τ_m, i.e. $b \neq 0$ in Eqn 5.10.

5.5 **TOCSY**

TOCSY (Fig. 5.8) is a popular alternative to COSY and DQF-COSY for large biological molecules. This is because it yields in-phase cross peaks which, unlike the antiphase cross peaks in COSY, do not tend to disappear as a result of the large linewidths encountered for molecules of high molecular mass. Another property of TOCSY is that it correlates remotely connected spins as well as directly connected spins and can thus be used to map out entire spin systems. The key feature of the TOCSY experiment is that it uses a period of *spin locking* to achieve coherence transfer.

TOCSY: Total Correlation Spectroscopy.

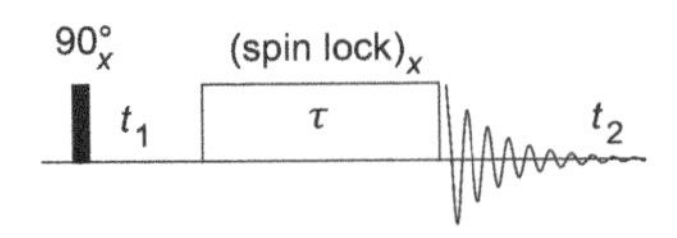

Fig. 5.8 The TOCSY pulse sequence.

In its simplest form, the spin locking field is just a long, strong radiofrequency pulse along a specified axis in the rotating frame. If $\omega_1=-\gamma B_1$ is much larger than all the offset frequencies Ω, precession around the z-axis is suppressed, all chemical shift differences become irrelevant, and the spins effectively become *equivalent*.

It needs a small amount of quantum mechanics to work out exactly how the various product operators evolve under the influence of the spin locking field, as explained in Chapter 10. The results for an IS spin system, with the spin locking field along the x-axis, are that the sum of the x-magnetizations of the two spins does not evolve, while their difference does:

$$I_x+S_x \xrightarrow{\text{spin lock}} I_x+S_x$$
$$I_x-S_x \xrightarrow{\text{spin lock}} (I_x-S_x)\cos(2\pi J_{IS}\tau)+(2I_yS_z-2I_zS_y)\sin(2\pi J_{IS}\tau). \quad (5.13)$$

These equations ignore the effects of relaxation during the relatively long spin-lock period. In particular cross relaxation will give rise to so-called *rotating frame Overhauser effects* (ROEs). The TOCSY experiment may be performed in a manner designed to maximize ROEs while minimizing the effects of evolution under the J-coupling; in this case it is referred to as ROESY.

The y and z components evolve in an analogous manner, but also precess around the x-axis at frequency ω_1. Note that the frequency of evolution of in-phase into antiphase terms under the influence of spin locking is $2\pi J_{IS}$ rather than the πJ_{IS} arising from free precession.

The evolution of the magnetization of a single spin, say I_x, can therefore be calculated as follows:

$$\begin{aligned}I_x &= \tfrac{1}{2}(I_x + S_x) + \tfrac{1}{2}(I_x - S_x) \\ &\xrightarrow{\text{spin lock}} I_x \tfrac{1}{2}\left[1+\cos(2\pi J_{IS}\tau)\right] + S_x \tfrac{1}{2}\left[1-\cos(2\pi J_{IS}\tau)\right] \\ &\quad + (2I_yS_z - 2I_zS_y)\tfrac{1}{2}\sin(2\pi J_{IS}\tau).\end{aligned} \tag{5.14}$$

Following only the spin I magnetization, the state of the IS spin system at the end of the t_1 period can be written:

$$\begin{aligned}I_z &\xrightarrow{90^\circ I_x} \xrightarrow{\Omega_I t_1 I_z} \xrightarrow{\pi J_{IS} t_1 2I_zS_z} \\ &-I_y \cos\Omega_I t_1 \cos(\pi J_{IS} t_1) + 2I_xS_z \cos\Omega_I t_1 \sin(\pi J_{IS} t_1) \\ &+ I_x \sin\Omega_I t_1 \cos(\pi J_{IS} t_1) + 2I_yS_z \sin\Omega_I t_1 \sin(\pi J_{IS} t_1).\end{aligned} \tag{5.15}$$

This process is analogous to the dephasing of transverse magnetization in an inhomogeneous B_0 field.

Now the first two of these terms will dephase during the spin locking if the B_1 field is spatially inhomogeneous (which it certainly will be) and along the same axis (x) as the initial 90° pulse. They can therefore be neglected. The fourth term, perhaps surprisingly, does not dephase completely and will give rise to an antiphase, dispersive contribution to the two-dimensional lineshapes. (This is because, if we also consider the spin S magnetization, there will also be a contribution from the product operator $2I_zS_y$ and the difference $2I_yS_z - 2I_zS_y$ is not affected by the spin-locking field, although the sum $2I_yS_z + 2I_zS_y$ is.) It is the third term that represents the desired component of the magnetization, as it will transfer into S_x during the spin locking:

See Section 10.5 in Part B for more details.

$$1+\cos 2A = 2\cos^2 A$$
$$1-\cos 2A = 2\sin^2 A$$

$$\begin{aligned}I_x \sin\Omega_I t_1 \cos(\pi J_{IS} t_1) &\xrightarrow{\text{spin lock}} \\ & I_x \sin\Omega_I t_1 \cos(\pi J_{IS} t_1)\cos^2(\pi J_{IS}\tau) \\ &+ S_x \sin\Omega_I t_1 \cos(\pi J_{IS} t_1)\sin^2(\pi J_{IS}\tau) \\ &+ (2I_yS_z - 2I_zS_y)\sin\Omega_I t_1 \cos(\pi J_{IS} t_1)\tfrac{1}{2}\sin(2\pi J_{IS}\tau).\end{aligned} \tag{5.16}$$

The antiphase term $(2I_yS_z - 2I_zS_y)$ can be suppressed by adding together several experiments with different values of the mixing time τ. This works because $\sin(2\pi J_{IS}\tau)$ changes sign as a function of τ while $\cos^2(\pi J_{IS}\tau)$ and $\sin^2(\pi J_{IS}\tau)$ do not. Both diagonal (I_x) and cross peaks (S_x) are in-phase in both dimensions and, as both have the same (amplitude) modulation, can both be phased to pure absorption (Fig. 5.9).

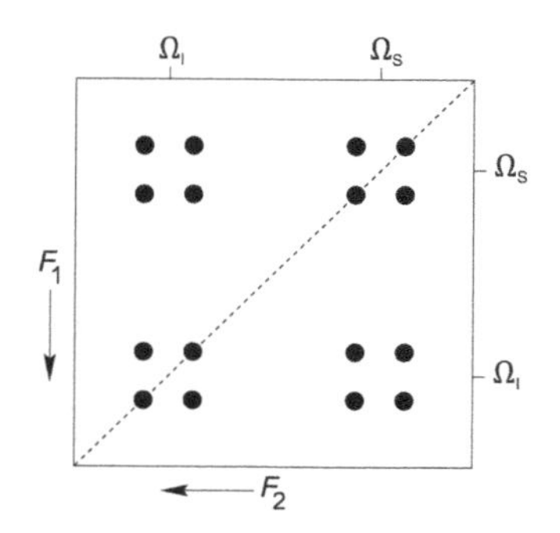

Fig. 5.9 Schematic TOCSY spectrum of an IS spin system, showing all peaks in positive absorption.

5.6 HMQC

All the two-dimensional NMR experiments considered so far have been homonuclear correlation techniques. As an example of a *heteronuclear* correlation

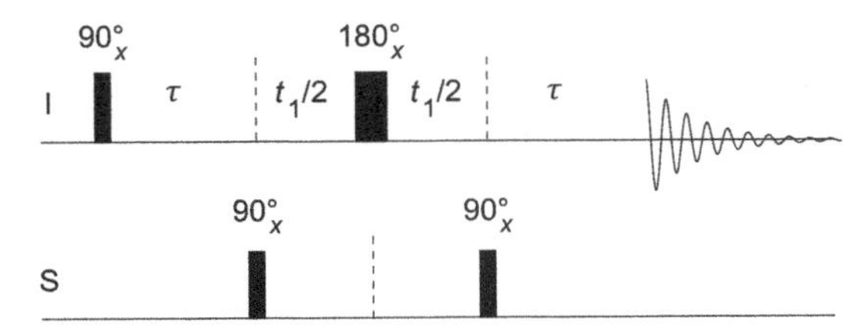

Fig. 5.10 The HMQC pulse sequence.

technique we will now perform a product operator analysis of the basic HMQC experiment (Fig. 5.10) for correlating protons and heteronuclei such as ^{13}C or ^{15}N through their J-couplings.

HMQC: Heteronuclear Multiple-Quantum Correlation.

The essence of a so-called inverse experiment is that sensitivity is maximized by observing the free induction decay of the more sensitive nucleus (typically ^{1}H). Starting with z-magnetization of spin I, we find:

$$\begin{aligned} & I_z \xrightarrow{90^\circ I_x} \xrightarrow{\Omega_I \tau I_z} \xrightarrow{\pi J_{IS} \tau 2 I_z S_z} \\ & -I_y \cos\Omega_I\tau \cos(\pi J_{IS}\tau) + 2I_x S_z \cos\Omega_I\tau \sin(\pi J_{IS}\tau) \\ & + I_x \sin\Omega_I\tau \cos(\pi J_{IS}\tau) + 2I_y S_z \sin\Omega_I\tau \sin(\pi J_{IS}\tau). \end{aligned} \tag{5.17}$$

The fixed free precession interval τ is set to $1/2J_{IS}$ so that the antiphase terms are maximized:

$$\begin{aligned} & = 2I_x S_z \cos\Omega_I\tau + 2I_y S_z \sin\Omega_I\tau \\ & \xrightarrow{90^\circ S_x} -2I_x S_y \cos\Omega_I\tau - 2I_y S_y \sin\Omega_I\tau. \end{aligned} \tag{5.18}$$

These two terms both represent mixtures of heteronuclear zero- and double-quantum coherence. In this experiment phase cycling is *not* used to separate the two. Note that the phase of these coherences depends on $\Omega_I\tau$ and that, at this point in the experiment, it is not clear how pure absorption lineshapes are going to be obtained. These heteronuclear multiple-quantum coherences now evolve during the t_1 period of the experiment. As a short cut in the calculation we can note that the 180° pulse on the I spin refocuses the I-spin chemical shifts during the evolution period. Therefore we need only calculate the evolution under the S-spin chemical shift together with the inversion of I_y by the 180° pulse:

Remember (Section 4.3) that zero- and double-quantum coherences do not evolve under the IS J-coupling.

$$\begin{aligned} & \xrightarrow{180^\circ I_x} -2I_x S_y \cos\Omega_I\tau + 2I_y S_y \sin\Omega_I\tau \\ & \xrightarrow{\Omega_S t_1 S_z} -2I_x S_y \cos\Omega_I\tau \cos\Omega_S t_1 + 2I_x S_x \cos\Omega_I\tau \sin\Omega_S t_1 \\ & \qquad + 2I_y S_y \sin\Omega_I\tau \cos\Omega_S t_1 - 2I_y S_x \sin\Omega_I\tau \sin\Omega_S t_1. \end{aligned} \tag{5.19}$$

Thus, the only evolution during t_1 occurs at the chemical shift frequency of spin S. The heteronuclear multiple-quantum coherences are now reconverted into I-spin single-quantum coherence:

$$\xrightarrow{90^\circ S_x} -2I_x S_z \cos\Omega_I\tau \cos\Omega_S t_1 + 2I_y S_z \sin\Omega_I\tau \cos\Omega_S t_1 \tag{5.20}$$

where we have retained only the terms that lead to observable signals. These antiphase coherences evolve under the influence of the spin I chemical shift and the coupling J_{IS} during the final interval prior to acquisition:

$$\begin{aligned}&\xrightarrow{\Omega_{\mathrm{I}}\tau I_z} -2I_xS_z\cos^2\Omega_{\mathrm{I}}\tau\cos\Omega_{\mathrm{S}}t_1 - 2I_yS_z\cos\Omega_{\mathrm{I}}\tau\sin\Omega_{\mathrm{I}}\tau\cos\Omega_{\mathrm{S}}t_1\\&\qquad + 2I_yS_z\cos\Omega_{\mathrm{I}}\tau\sin\Omega_{\mathrm{I}}\tau\cos\Omega_{\mathrm{S}}t_1 - 2I_xS_z\sin^2\Omega_{\mathrm{I}}\tau\cos\Omega_{\mathrm{S}}t_1\\&= -2I_xS_z\cos\Omega_{\mathrm{S}}t_1\\&\xrightarrow{\pi J_{\mathrm{IS}}\tau 2I_zS_z} -I_y\cos\Omega_{\mathrm{S}}t_1.\end{aligned} \tag{5.21}$$

The chemical shift evolution during the first τ period has been refocused in the second. Thus pure absorption lineshapes can be obtained. The final multiplet structure of the peaks in the two-dimensional spectrum is a singlet in F_1 and an in-phase doublet in F_2. Broadband decoupling of the S spin can be used to obtain singlets in F_2. Note that in this experiment, as in other heteronuclear correlation techniques, there are no diagonal peaks and only one cross peak per pair of J-coupled spins.

5.7 HSQC

HSQC: Heteronuclear Single-Quantum Correlation.

As a method of J-correlating heteronuclei, HMQC has largely been superseded by HSQC. This technique, which turns out to be much simpler to understand, is the basis of many more complex *three-dimensional* NMR experiments which play a central role in modern NMR studies of biomolecules such as proteins.

A simple HSQC pulse sequence is shown in Fig. 5.11. To break the monotony of analysing pulse sequences step-by-step, we treat this experiment as being built up from a small number of basic, and by now familiar elements. Knowing the behaviour of each of the building blocks, we can combine them in order to understand the whole sequence. This approach is almost essential when designing more complex experiments, which are developed by combining standard elements in interesting ways.

The first element of the HSQC experiment should be familiar as an INEPT sequence, which we have already analysed in considerable detail (see Section 4.2). At the end of this block the state of a two-spin IS system is (Eqn 4.3)

$$a2I_zS_y + bS_y, \tag{5.22}$$

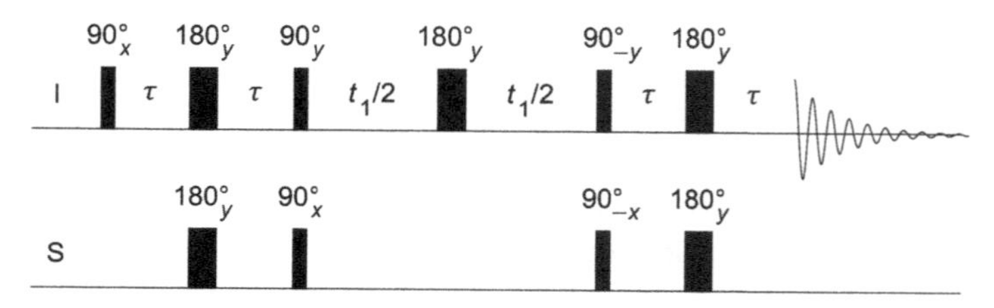

Fig. 5.11 The HSQC pulse sequence.

where a and b are the inherent amplitudes of the I- and S-spin magnetizations, respectively. As shown in Fig. 4.2, the bS_y contribution is normally removed by phase cycling and we will assume below that this has been carried out. The INEPT block is followed by a t_1 evolution period, in the middle of which a 180° pulse is applied to spin I. The evolution arising from the J-coupling of I and S will be refocused, leaving only evolution under the S-spin chemical shift together with the effects of the 180° pulse on spin I. Thus, at the end of the evolution period, the state of the system will be

$$-a2I_zS_y\cos\Omega_St_1 + a2I_zS_x\sin\Omega_St_1. \qquad (5.23)$$

The final element of HSQC is simply a reversal of the initial INEPT sequence (i.e. all the pulses and delays applied in reverse order, with the phase of the 90° pulses inverted), except that the first pulse has been removed. Taking these changes into account, we can deduce that the sequence will effect the transformation

$$2I_zS_y \rightarrow I_y. \qquad (5.24)$$

Thus the evolution of the first term in Eqn 5.23 can be easily calculated. The pulse sequence is followed by observation of the NMR signal of spin I, so that only this first term will give rise to any observable signal. The remaining term in Eqn 5.23 is slightly more complicated but does not result in an observable I-spin signal.

The second term in Eqn 5.23 is converted into (unobservable) zero- and double-quantum coherences.

Therefore, the observable I-spin magnetization at the end of the HSQC pulse sequence is

$$-aI_y\cos\Omega_St_1. \qquad (5.25)$$

As with HMQC (Eqn 5.21), the signal from the I spin has been labelled with a modulation arising from the S spin (typically ^{13}C or ^{15}N), and then returned to I (typically ^1H) for detection. Modifying the relative phases of the INEPT and reverse INEPT sequences allows detection of a sine-modulated component, and so the sign of Ω_S can be determined (see Section 2.4).

5.8 Three-dimensional NMR

NMR is routinely used to study large proteins and other biomolecules, and for many of these systems two-dimensional NMR techniques do not provide enough information. It is common to use three-dimensional NMR experiments, which allow three different frequencies to be correlated. There is a huge variety of such experiments, and it would take most of this book to analyse even the most common ones in detail. Fortunately, however, it is not necessary to perform such an analysis, as many of them can be easily understood by combining the basic building blocks we have already seen.

See, for example, Cavanagh *et al.* *Protein NMR Spectroscopy* (2006) and Kay *et al.* *Three-dimensional Triple Resonance NMR Spectroscopy of Isotopically Enriched Proteins* (1990).

One important group of experiments combines an HSQC sequence with a homonuclear technique such as NOESY or TOCSY. This can be achieved by replacing the first 90° pulse of the homonuclear experiment with an HSQC sequence to make the HSQC-NOESY and HSQC-TOCSY experiments. Alternatively the first

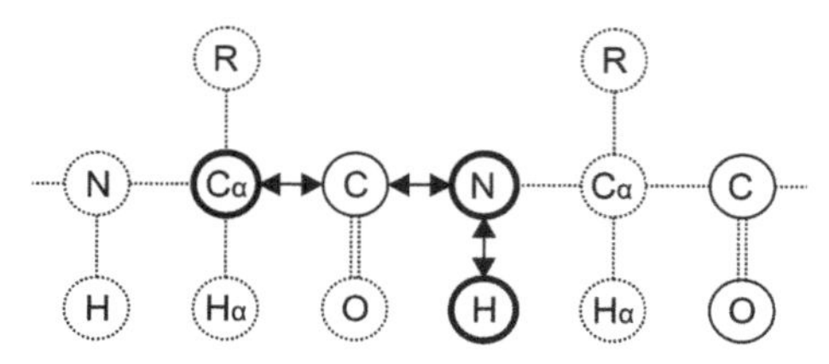

Fig. 5.12 The sequence of transfers during the HN(CO)CA pulse sequence. A fragment of the peptide backbone is shown schematically. Bold circles indicate spins whose frequencies are recorded in the spectrum.

pulse of the HSQC sequence can be replaced with the homonuclear technique, giving the NOESY-HSQC and TOCSY-HSQC experiments.

Similarly, in an era when many proteins can be uniformly labelled with both ^{13}C and ^{15}N, it is possible to correlate the frequencies of three different heteronuclei (e.g. ^{1}H, ^{15}N, and ^{13}C nuclei) by nesting one HSQC sequence inside another. Two very important examples of this approach are the complementary three-dimensional experiments HNCA and HN(CO)CA. In the former, the chemical shift of an amide ^{15}N nucleus in an amino acid in the protein backbone is correlated with the ^{13}C chemical shift of a Cα carbon in the same amino acid residue and in the preceding residue. In contrast, in the HN(CO)CA experiment, the amide nitrogen shift is correlated with the Cα shift of the preceding residue alone. By comparing HNCA and HN(CO)CA spectra, it is possible to achieve a complete assignment of the amide and Cα resonances.

In a break from the usual tedious acronyms, note that the pulse sequence names HNCA and HN(CO)CA are intended to be self-descriptive.

The HNCA and HN(CO)CA experiments feature multiple 'out-and-back' INEPT transfers, from ^{1}H to ^{15}N (and vice versa) and from ^{15}N to ^{13}C (and vice versa). Another characteristic feature of the two pulse sequences is that the carbonyl (CO) carbons are manipulated separately from the Cα carbons using frequency-selective ^{13}C pulses: as the signals from CO and Cα carbons occupy two separate (non-overlapping) parts of the spectrum, it is possible to, in effect, treat the two sorts of ^{13}C as different nuclides. In the HN(CO)CA experiment (Fig. 5.12) INEPT transfers occur from ^{1}H to ^{15}N, then from ^{15}N to ^{13}CO, and finally from ^{13}CO to ^{13}Cα, before reversing the whole process. The magnetization is not allowed to evolve at the ^{13}CO frequency, but is instead immediately transferred to the next spin, and so only three separate frequency dimensions are recorded.

5.9 Summary

- Many simple two-dimensional NMR experiments can be basically understood by analysing coupled pairs of spins using product operators.
- Results for multi-spin systems are usually analogous, but detailed calculations can also reveal undesirable terms.
- In two-dimensional NMR spectra it is not always possible to phase all the peaks into absorption and it may be necessary to sacrifice the appearance of the diagonal to optimize the cross peaks.

- The same pulse sequence with different phase cycling can give quite different results.
- Phase cycling to select population terms also allows zero-quantum coherences through the filtration step.
- COSY, DQF-COSY, NOESY, and TOCSY all allow correlations between homonuclear spin pairs (usually two ^{1}H nuclei) to be studied.
- HMQC and HSQC allow correlations between ^{13}C or ^{15}N nuclei and directly bonded ^{1}H nuclei to be studied.
- Two-dimensional NMR experiments can be 'nested' to give three-dimensional NMR experiments.

5.10 Exercises

Worked solutions to the exercises are available on the Online Resource Centre at www.oxfordtextbooks.co.uk/orc/hore2e/

5.1. Performing the COSY experiment with two 90°_x pulses as in Fig. 5.1 yields diagonal peak signals that are sine modulated as a function of t_1 and cross-peak signals that are cosine modulated (see Eqn 5.2). Show that changing the first pulse to 90°_y results in diagonal peak signals that are cosine modulated as a function of t_1 and cross-peak signals that are sine modulated. (Note that the signals from these two experiments can be used as the inputs to a *hypercomplex Fourier transformation*, as described in Section 2.4.)

5.2. Eqn 5.3 shows eight product operators present at the start of the t_2 (acquisition) period of a COSY experiment performed on a three-spin ISR system. Write down the 16 product operators you would expect to find at the start of the t_2 period of a COSY experiment performed on a four-spin ISQR system. (There is no need to give the signs, associated modulations, or to perform the complete product operator calculation.)

5.3. A cosine modulated signal is obtained as a function of t_1 for the HMQC pulse sequence in Fig. 5.10 (see Eqn 5.21). How might the pulse sequence be modified to obtain a sine modulated signal?

5.4. The HMQC experiment is normally performed to correlate I and S spins through their one-bond J-couplings, $^{1}J_{IS}$. The Heteronuclear Multiple-Bond Correlation (HMBC) experiment is a version of HMQC that allows correlation through multiple-bond J-couplings: $^{2}J_{IS}$, $^{3}J_{IS}$, $^{4}J_{IS}$, etc. Suggest the simplest modification of HMQC that allows such correlations to be observed.

5.5. In the early years of two-dimensional NMR, heteronuclear correlation experiments were usually carried out by having ^{1}H magnetization evolve during the t_1 period and detecting the ^{13}C or ^{15}N signal directly in the t_2 period. However, this approach is much less sensitive than that used in the so-called inverse experiments HMQC and HSQC and has fallen out of favour. Despite this overwhelming sensitivity disadvantage, suggest some *advantages* of the direct detection approach.

6 Phase cycling and pulsed field gradients

6.1 Introduction

We have already encountered the concept of *phase cycling*: NMR experiments can be repeated several times with different pulse phases and the resulting signals added and subtracted in such a way as to cancel any unwanted contributions. For example, a two-step cycle was used in the INEPT experiment (Section 4.2) to select the enhanced S-spin magnetization and to reject the unenhanced component; and four steps were used to ensure the correct operation of the double-quantum filter (Section 4.3), cancelling any single-quantum magnetization present during the filter delay Δ. We have also noted in passing (Section 5.4) that the pulse sequences for the DQF-COSY and NOESY two-dimensional experiments are essentially identical, both consisting of three radiofrequency pulses, and that it is principally the phase cycling that distinguishes them.

An alternative to phase cycling is to use *pulsed field gradients* (i.e. spatially inhomogeneous magnetic fields that can be turned on or off at will) to dephase and later rephase only the required coherences. This method has largely supplanted phase cycling in modern NMR spectroscopy of liquid samples (although some additional phase cycling is nearly always required). In contrast, for technical reasons, phase cycling remains the preferred method in NMR of solid samples.

Phase cycling and/or pulsed field gradients are essential parts of any modern NMR experiment; the purpose of this chapter is to describe how they work and how they can be used to select only the desired signals.

6.2 Coherence transfer pathways

Every product operator can be assigned one or more *coherence orders*. Pure single-, double-, and triple-quantum operators have coherence orders $p=\pm1$, ±2, and ±3, respectively, while population and zero-quantum operators have $p=0$. Some examples are given in Table 6.1. Thus, a two-spin product operator such as $2I_xS_y$ has coherence orders $p=\pm2$ and 0 as it represents a sum of double- and zero-quantum coherences. Note that the coherence order is a signed quantity; the meaning of this will become apparent below.

Table 6.1 Examples of some product operators and their coherence orders, *p*.

Operators	*p*
S_z, $2I_zS_z$, $4I_zS_zR_z$, $(I_xS_x+I_yS_y)$, $(2I_zS_yR_x-2I_zS_xR_y)$	0
I_x, I_y, $2I_zS_y$, $4I_xS_zR_z$	±1
$(I_xS_x-I_yS_y)$, $(2I_zS_xR_y+2I_zS_yR_x)$	±2

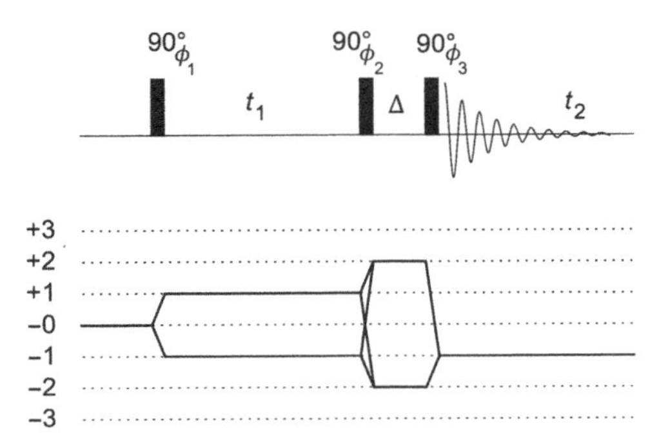

Fig. 6.1 Pulse sequence and coherence transfer pathway diagram for the double-quantum filtered COSY experiment.

During the free precession intervals of an NMR experiment the spectroscopist must have control over which coherence orders are present. For example, during the t_1 period of a DQF-COSY experiment one wishes to have only single-quantum ($p=\pm1$) coherences evolving, but during the very brief delay that follows, only double-quantum ($p=\pm2$) coherences are wanted. For every NMR experiment we can draw a *coherence level diagram* or *coherence transfer pathway diagram* which shows the *desired* coherence orders present in each free precession interval of the pulse sequence. The coherence transfer pathway diagram for a DQF-COSY experiment, for example, is shown in Fig. 6.1.

The following features—general to all NMR experiments—should be noted:

The coherence transfer pathway starts at $p=0$. The relatively long relaxation delay that precedes the pulse sequence should ensure that the spin system is initially at thermal equilibrium, with only z-magnetization present.

The coherence transfer pathway finishes at $p=-1$. Observable magnetization is always single-quantum coherence ($p=\pm1$) and, if quadrature detection is used (which it ought to be), the sense of precession of this magnetization will be known. The two coherence orders $p=+1$ and $p=-1$ correspond to two apparently counter-rotating components of the single-quantum coherence in the rotating frame and quadrature detection will detect only one of these. The usual convention is that quadrature detection takes the x component of the precessing magnetization as the real signal and the y component as the imaginary signal (Section 2.2) and this corresponds to detection of the $p=-1$ coherence.

See Ernst, Bodenhausen, and Wokaun, *Principles of Nuclear Magnetic Resonance in One and Two Dimensions* (1985).

During the free precession intervals (t_1 and Δ in this case) prior to the acquisition period *both $p=+n$ and $p=-n$ coherences are shown as present.* When free precession is interrupted by a 90° pulse we get an *amplitude-modulated* signal (see Chapter 2) which can be thought of as the sum of two counter-rotating components in the rotating frame; it is these that are represented by (for example) the pathways with $p=+1$ and -1 during the evolution period t_1.

The coherence order does not change during free precession. Only pulses can change the coherence order. A 90° pulse will in general change a coherence order p into all possible coherence orders allowed within the spin system; a perfect 180° pulse simply changes a coherence order p into $-p$.

The coherence transfer pathway diagram only shows the desired coherence transfer pathways during the pulse sequence. In DQF-COSY for example, if the first pulse is not precisely 90° then some z-magnetization ($p=0$) will be left during the t_1 period. Similarly, the second 90° pulse will excite $p=0, \pm 1$, (and if the spin system is big enough), $\pm 3, \pm 4$, etc. All such extraneous coherence orders, which must be suppressed by phase cycling and/or pulsed field gradients, are excluded from the diagram.

The great beauty of coherence transfer pathway diagrams is that they give immediate insight into what is actually going on during a pulse sequence. As we shall see, they also allow us to arrive quickly at a simple recipe for cycling phases or inserting pulsed field gradients.

6.3 Phase cycling

What one might term the *First Rule of Phase Cycling* can be expressed as follows:

> If the phase of a pulse or a group of pulses is shifted by ϕ, then a coherence transfer pathway undergoing a change in coherence order Δp experiences a phase shift of $-\phi\Delta p$ in the detection period.

The desired change in coherence order Δp for any given pulse can be found simply by inspection of the coherence transfer pathway diagram. The corresponding pathway can then be selected by making the receiver phase ϕ_{Rx} follow the phase of the coherence pathway:

$$\phi_{Rx} = -\phi\Delta p. \tag{6.1}$$

Other coherence transfer pathways will experience different phase shifts and should sum to zero over the phase cycle. For example, in DQF-COSY (Fig. 6.1) the final 90° pulse must change the coherence order by $\Delta p=-3$ (i.e. $+2\rightarrow-1$) *and* by $\Delta p=+1$ (i.e. $-2\rightarrow-1$). If the phase of the pulse is cycled through the values

$$\phi_3 = 0°, 90°, 180°, 270° \tag{6.2}$$

then, taking the pathway $\Delta p = -3$ first, we find that the receiver phase should go as

$$\phi_{Rx} = +3\phi_3 = 0°, 270°, 540°, 810° \tag{6.3}$$

which is equivalent to

$$\phi_{Rx} = 0°, 270°, 180°, 90°. \tag{6.4}$$

On the other hand, if the pathway $\Delta p=+1$ is considered then the receiver phase cycle is

$$\phi_{Rx} = -\phi = 0°, -90°, -180°, -270° \tag{6.5}$$

which is easily seen to be identical to Eqn 6.4. This simple four-step phase cycle thus selects both the pathways $\Delta p=-3$ and $+1$ as desired.

But how did we know that we had to use a *four*-step phase cycle (i.e. increment the pulse phase ϕ by 90° each time)? Why not three, five, or six steps? And, although the above rule tells us whether particular pathways survive, how can we be sure that the undesirable pathways are being suppressed? The answers to these questions are provided by the *Second Rule of Phase Cycling*:

> If a phase cycle uses steps of 360°/N then, along with the desired pathway Δp, pathways $\Delta p \pm nN$, where $n=1, 2, 3, \ldots$, will also be selected. All other pathways are suppressed.

Table 6.2 Possible changes in coherence order Δp without (A) and with (B) phase cycling of the final pulse in the DQF-COSY experiment.

A	B
...	...
+7	
+6	
+5	+5
+4	
+3	
+2	
+1	+1
0	
−1	
−2	
−3	−3
−4	
−5	
−6	
−7	−7
...	...

One way to use this rule is to write out all possible changes in the coherence order that can occur as a result of the pulse (Table 6.2, column A). In our DQF-COSY example we want to cycle the phase of the third pulse to select the pathways $\Delta p=-3$ and $+1$, and suppress all intermediate ones ($\Delta p=-2, -1$, and 0). According to the above rule, to do this we must have $N=4$. Deleting the pathways suppressed by the phase cycle we obtain column B in Table 6.2. Thus our phase cycle also retains the pathways $\Delta p=-7$ and $+5$, i.e. it selects six- as well as two-quantum coherences during the delay. However, these will make only vanishing contributions to the final signal; even if the spin system is large enough to support such high-order coherences, their excitation will be very inefficient.

Similarly, we can consider cycling the phase of the first pulse ϕ_1 in the DQF-COSY experiment (Fig. 6.1). From the Second Rule, we can see that two steps are needed to select $\Delta p=+1$ *and* $\Delta p=-1$ simultaneously, rejecting the unwanted $\Delta p=0$. These two steps are

$$\phi_1 = 0°, 180° \tag{6.6}$$

and the First Rule gives the corresponding receiver phase cycle as

$$\phi_{Rx} = 0°, 180°. \tag{6.7}$$

In the context of two-dimensional NMR, this simple phase cycling of the first pulse is known as *axial peak suppression* as it prevents any coherence pathways that have $p=0$ in the t_1 period contributing to the final signal, where they would produce peaks at $F_1=0$ (so-called *axial peaks*) in the two-dimensional spectrum.

All coherence transfer pathways start with coherence order $p=0$ and finish with coherence order $p=-1$. Therefore, if there are N pulses in a pulse sequence then only N–1 of these pulses need to be phase cycled to select the desired coherence transfer pathways. We have already derived phase cycles for the phases of the first and third pulses in the DQF-COSY experiment and so we have the raw materials for a complete phase cycle. But at the moment we still have two independent phase cycles: we now need to *nest* these (like loops in a computer program) to give us a single phase cycle for the whole experiment. For each of the two steps in the phase cycle of ϕ_1 we must perform a complete phase cycle for ϕ_3 (four steps) yielding the complete eight-step phase cycle for the experiment in Fig. 6.1:

$$\begin{aligned}
\phi_1 &= 0^\circ \quad 0^\circ \quad 0^\circ \quad 0^\circ \quad 180^\circ \quad 180^\circ \quad 180^\circ \quad 180^\circ \\
\phi_2 &= 0^\circ \quad 0^\circ \quad 0^\circ \quad 0^\circ \quad 0^\circ \quad 0^\circ \quad 0^\circ \quad 0^\circ \\
\phi_3 &= 0^\circ \quad 90^\circ \quad 180^\circ \quad 270^\circ \quad 0^\circ \quad 90^\circ \quad 180^\circ \quad 270^\circ \\
\phi_{Rx} &= 0^\circ \quad 270^\circ \quad 180^\circ \quad 90^\circ \quad 180^\circ \quad 90^\circ \quad 0^\circ \quad 270^\circ
\end{aligned} \tag{6.8}$$

It can readily be seen that if we had phase cycled ϕ_1 for each step in the phase cycle of ϕ_3, we would have ended up with an identical phase cycle but performed in a different order (and the order in which a phase cycle is performed should be irrelevant).

But why did we choose ϕ_1 and ϕ_3 as the two pulses to phase cycle in the DQF-COSY example? Why not ϕ_2 and ϕ_3 or ϕ_1 and ϕ_2? These questions reveal that, despite the rules presented above, there is still something of an art to phase cycling and experience is often useful. In fact, cycling of ϕ_2 and ϕ_3 in Fig. 6.1 yields a phase cycle identical to that presented in Eqn 6.8 except that it is again performed in a different order, whereas cycling of ϕ_1 and ϕ_2 alone fails as it cannot suppress coherence pathways that have $p=0$ during the Δ period. This is explored further in the exercises.

Clearly the phase cycling of an NMR experiment with several cycles nested within cycles can end up being quite time-consuming; this is where pulsed field gradients come in.

6.4 Pulsed field gradients

The idea behind the use of magnetic field gradients for coherence pathway selection is that:

> If a static magnetic field is applied briefly along the z-axis, the phase shift experienced by a coherence will be proportional to its order, p.

The similarity to the First Rule of Phase Cycling should be clear. To be more specific, if a field of strength B_G is switched on for a time τ, the phase shift (in radians) is $p\gamma B_G\tau$, where γ is the magnetogyric ratio of the nucleus in question. If the field varies in strength across the NMR sample (which is what is meant by a field *gradient*), then coherences in different regions of the NMR tube experience different phase shifts. If the field gradient is strong enough and lasts long enough, coherences with non-zero orders (i.e. with $p\neq0$) will be completely dephased by the inhomogeneous field. A second pulsed field gradient, applied later in the pulse sequence, with suitable B_G and τ, can selectively undo the effect of the first, and so pick out the required change in coherence order. An example should clarify how this works.

Suppose we want to select the pathway $+2\rightarrow-1$ at the end of a DQF-COSY sequence. If we put two pulsed field gradient pulses, labelled G_1 and G_2, either side of the 90° pulse, as in Fig. 6.2(a), the total phase shift of the desired pathway will be

$$\Phi = +2\gamma B_{G_1}\tau_1 - \gamma B_{G_2}\tau_2. \tag{6.9}$$

By making the second gradient twice as strong or twice as long as the first, we ensure that $\Phi=0$ and only the required pathway is rephased. Other pathways suffer non-zero net phase shifts and so remain dephased. Similarly, if the $-2\rightarrow-1$ pathway is required, all that is needed is to reverse the direction of one of the gradients, as shown in Fig. 6.2(b). Evidently, one cannot select *both* pathways simultaneously, in contrast to phase cycling. As a consequence, there is often a signal-to-noise penalty associated with the use of field gradients. Coherence transfer pathways in heteronuclear experiments may be selected in exactly the same way, except that the appropriate magnetogyric ratios must be included in expressions analogous to Eqn 6.9.

A 'gradient-selected' DQF-COSY pulse sequence and its coherence transfer pathway diagram are shown in Fig. 6.3. As described above, the final pulse is bracketed by two pulsed field gradients, $G_2=-2G_1$ (i.e. $B_{G_2}\tau_2=-2B_{G_1}\tau_1$). Acquisition starts immediately after G_2. Note that by not putting a gradient pulse into the t_1 period, it is still possible to use the method outlined in Section 2.4 (or similar) to obtain pure two-dimensional absorption lineshapes with sign discrimination in F_1.

Pulsed field gradients have advantages over phase cycling that more than compensate for the modest loss in sensitivity noted above. Because the desired coherence transfer pathway can usually be selected in a single experiment, pulsed field gradients put a much smaller demand on the stability of the spectrometer than does phase cycling, where the unwanted signals are only cancelled by subtraction at the end of a lengthy set of repetitions of the pulse sequence. Furthermore, field gradients allow one to arrange the experiment so as to obtain the desired resolution and sensitivity, without being constrained to complete a prolonged phase cycle for each t_1 value in a two-dimensional experiment. However, as noted in Section 6.1, some additional phase cycling is often still performed. For example, in the DQF-COSY pulse sequence in Fig. 6.3, one would probably still choose to perform axial peak suppression using a two-step phase cycle of the first pulse, as described in Section 6.3.

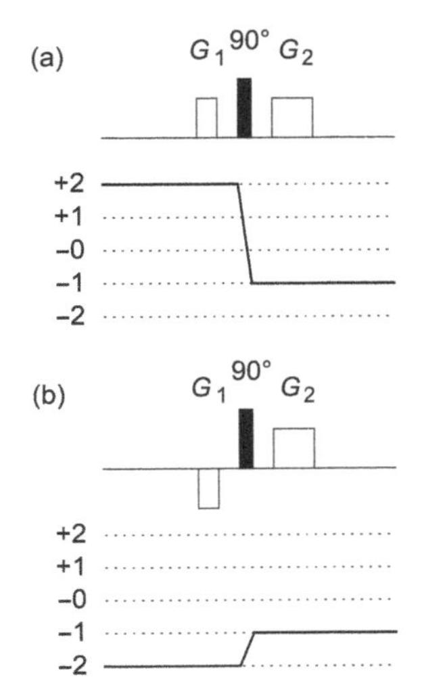

Fig. 6.2 Use of pulsed field gradients to select the different coherence transfer pathways: (a) $p=+2\rightarrow p=-1$, and (b) $p=-2\rightarrow p=-1$.

In practice, extra steps need to be taken to refocus the phase errors that arise from evolution during Δ, which has had to be increased in duration to accommodate the first pulsed field gradient.

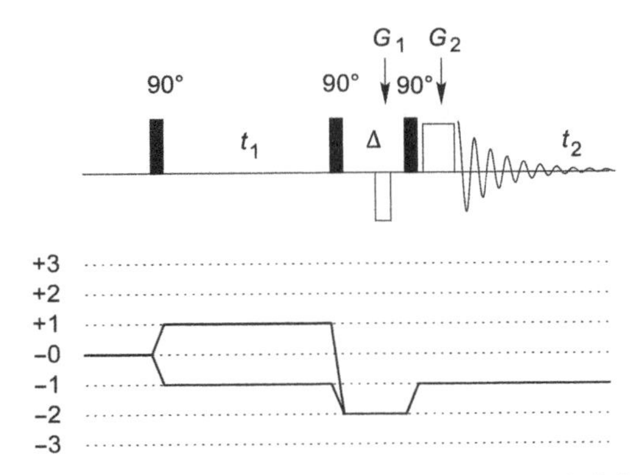

Fig. 6.3 DQF-COSY pulse sequence with pulsed field gradients instead of phase cycling.

6.5 **Summary**

- NMR experiments can be understood in terms of coherence transfer pathways.
- Using two simple rules, phase cycles can be designed to select for particular coherence pathways.
- Phase cycles for individual pulses are nested together to give the complete phase cycle for the pulse sequence.
- Nested phase cycles are rarely unique; often there are several alternatives that are equally successful.
- Pulsed magnetic field gradients can be used to select coherence transfer pathways.
- Field gradients are usually more effective than phase cycling at suppressing undesired terms, but this comes with a small cost in sensitivity.
- Field gradients and phase cycling are often used together.

6.6 **Exercises**

Worked solutions to the exercises are available on the Online Resource Centre at www.oxfordtextbooks.co.uk/orc/hore2e/

6.1. Draw coherence transfer pathway diagrams for (a) the COSY experiment (Fig. 5.1), (b) the NOESY experiment (Fig. 5.7), and (c) the HMQC experiment (Fig. 5.10). (For the heteronuclear HMQC experiment, remember that you will need to draw separate coherence transfer pathway diagrams for the I and S spins.)

6.2. Remembering that, in experimental practice, the flip angles of the pulses may deviate slightly from 90°, design a two-step phase cycle for the COSY experiment (Fig. 5.1).

6.3. The NOESY experiment (Fig. 5.7) requires a two-step phase cycle to suppress axial peaks and a four-step phase cycle to select $p=0$ during τ_m. Write down these two phase cycles separately and then nest them to give the final eight-step NOESY phase cycle.

6.4. Confirm that cycling the phases ϕ_2 and ϕ_3 in the DQF-COSY experiment (Fig. 6.1) yields the same eight-step phase cycle (but performed in a different order) as that obtained by cycling phases ϕ_1 and ϕ_3 (Eqn 6.8), as described in Section 6.3.

6.5. Confirm that cycling the phases ϕ_1 and ϕ_2 in the DQF-COSY experiment (Fig. 6.1) alone cannot suppress $p=0$ during the Δ delay, as described in Section 6.3.

6.6. A strong pulsed field gradient is often applied before the very first radiofrequency pulse is applied in a pulse sequence. Suggest why this is beneficial.

Part B

Quantum Mechanics

7 Quantum mechanics

7.1 Introduction

In Part B of this book we turn to a rigorous quantum mechanical description of NMR experiments. We explain what product operators really are, justify the way they have been used to understand the operation of NMR pulse sequences, and derive many of the results that were quoted without proof in Part A. We begin with the quantum mechanical description of a two-level system: this may seem very abstract, but we soon come to more concrete examples.

Throughout the next four chapters we assume a passing familiarity with basic quantum mechanics. Excellent general textbooks include Atkins and Friedman *Molecular Quantum Mechanics* (2010), while a more focused treatment of two-level systems can be found in Jones and Jaksch *Quantum Information, Computation and Communication* (2012).

7.2 Ket and bra vectors

The most convenient notation for describing two-level systems, such as spin-$\frac{1}{2}$ nuclei in magnetic fields, is the bra and ket notation developed by Dirac. We will begin by reviewing some basic properties of this notation, before introducing spins in Section 7.4. Any arbitrary state of a two-level system is specified by a ket $|\psi\rangle$, which can in general be written as a linear combination, or *superposition*, of two basis states, one for each level:

See Jones and Jaksch (2012) for a more detailed introduction to Dirac's bra(c)ket notation.

$$|\psi\rangle = c_1|1\rangle + c_2|2\rangle. \tag{7.1}$$

The basis states $|1\rangle$ and $|2\rangle$ can be thought of as orthogonal unit vectors, while the coefficients c_1 and c_2 measure the contributions of these two vectors to the superposition. Unlike conventional vectors, however, c_1 and c_2 can be complex numbers, and so do not simply describe the amounts of $|1\rangle$ and $|2\rangle$ in the superposition, but also the relative phase with which they add together.

Some important properties of vectors and matrices are summarized in Appendix B.

Following this vector approach, it is often convenient to write a ket as a column vector, whose elements are the coefficients in Eqn 7.1:

$$\boldsymbol{\psi} = \begin{pmatrix} c_1 \\ c_2 \end{pmatrix}. \tag{7.2}$$

As in Part A, vectors such as $\boldsymbol{\psi}$ are set in ***bold italic*** type. We will switch back and forth between the $|\psi\rangle$ and $\boldsymbol{\psi}$ notations, using whichever is more convenient for our purposes.

Just as for conventional vectors we can define a scalar product, or *inner product*, but it is necessary to allow for the fact that c_1 and c_2 can be complex. For each ket there is a corresponding bra $\langle\psi|$ defined by

$$\langle\psi| = \langle 1|c_1^* + \langle 2|c_2^*. \tag{7.3}$$

This can be written as a row vector

The asterisk indicates the complex conjugate.

$$\boldsymbol{\psi}^{\dagger} = \begin{pmatrix} c_1^* & c_2^* \end{pmatrix} \tag{7.4}$$

where † indicates the matrix adjoint (the complex conjugate of the transpose of the matrix). The scalar product of $\langle\psi|$ and $|\psi\rangle$ is then given by

As the coefficients c_1 and c_2 are simply numbers (scalars) they can be moved within this equation at will.

$$\begin{aligned}\langle\psi|\psi\rangle &= \left[\langle 1|c_1^* + \langle 2|c_2^*\right]\left[c_1|1\rangle + c_2|2\rangle\right] \\ &= c_1^*c_1\langle 1|1\rangle + c_1^*c_2\langle 1|2\rangle + c_2^*c_1\langle 2|1\rangle + c_2^*c_2\langle 2|2\rangle.\end{aligned} \tag{7.5}$$

As $|1\rangle$ and $|2\rangle$ are orthogonal unit vectors, their scalar products are

Orthogonal unit vectors are said to be *orthonormal*.

$$\langle 1|1\rangle = \langle 2|2\rangle = 1 \qquad \langle 1|2\rangle = \langle 2|1\rangle = 0 \tag{7.6}$$

and so

The use of complex conjugates in defining the bra vector ensures that the lengths of vectors will be positive real numbers.

$$\langle\psi|\psi\rangle = c_1^*c_1 + c_2^*c_2 = |c_1|^2 + |c_2|^2, \tag{7.7}$$

which is the square of the length of $|\psi\rangle$. For convenience, $|\psi\rangle$ is usually *normalized* so that its length is equal to one.

This definition is not rigorously correct as the matrix product gives a matrix, not a scalar, but this can be side stepped by taking the matrix trace as discussed in Appendix B.

This result can also be derived directly from the vector representations in Eqns 7.2 and 7.4: the scalar product is then just a conventional matrix product

$$\langle\psi|\psi\rangle = \boldsymbol{\psi}^{\dagger}\boldsymbol{\psi} = \begin{pmatrix} c_1^* & c_2^* \end{pmatrix}\begin{pmatrix} c_1 \\ c_2 \end{pmatrix} = c_1^*c_1 + c_2^*c_2. \tag{7.8}$$

7.3 **Operators**

Operators are described in more detail in Appendix C; they are conventionally written with 'hats' on, as shown here.

An *operator* $\hat{A}$ acts on a ket $|\psi\rangle$ to give some new ket $|\psi'\rangle$. In general, the effect of $\hat{A}$ on $|\psi\rangle$ can be described as some combination of rotation and scaling. In some cases, however, $\hat{A}$ will simply scale (i.e. alter the length of) $|\psi\rangle$ without affecting its direction:

$$\hat{A}|\psi\rangle = a|\psi\rangle. \tag{7.9}$$

In this case, $|\psi\rangle$ is said to be an *eigenvector* of $\hat{A}$, with *eigenvalue a*.

If the operator $\hat{A}$ corresponds to an observable quantity, such as energy or angular momentum, then the mean value obtained from an experimental measurement of this quantity for a system in the state $|\psi\rangle$ is given by the *expectation value* of $\hat{A}$, which is simply the scalar product of $\langle\psi|$ with $\hat{A}|\psi\rangle$:

This expression assumes that $|\psi\rangle$ is normalized so that $\langle\psi|\psi\rangle = 1$.

$$\langle\hat{A}\rangle = \langle\psi|\hat{A}|\psi\rangle. \tag{7.10}$$

Suppose that the basis vectors, $|1\rangle$ and $|2\rangle$, are eigenvectors of $\hat{A}$, i.e.

$$\hat{A}|1\rangle = a_1|1\rangle, \qquad \hat{A}|2\rangle = a_2|2\rangle. \tag{7.11}$$

The effect of $\hat{A}$ on $|\psi\rangle$ is then

This uses the fact that the operator $\hat{A}$ is linear; see Appendix C.

$$\hat{A}|\psi\rangle = \hat{A}c_1|1\rangle + \hat{A}c_2|2\rangle = a_1c_1|1\rangle + a_2c_2|2\rangle \tag{7.12}$$

and so

Compare this expression with Eqns 7.5 and 7.7.

$$\begin{aligned}\langle\hat{A}\rangle &= \langle\psi|\hat{A}|\psi\rangle = \left[c_1^*\langle 1| + c_2^*\langle 2|\right]\left[a_1c_1|1\rangle + a_2c_2|2\rangle\right] \\ &= a_1|c_1|^2 + a_2|c_2|^2.\end{aligned} \tag{7.13}$$

The mean value obtained from an experiment is thus a weighted average of the two eigenvalues, a_1 and a_2. Any single experiment will result in one of these two values being measured, with probabilities $|c_1|^2$ and $|c_2|^2$; these are the probabilities that the system will be found in the corresponding basis states when a measurement is made.

In general, $|1\rangle$ and $|2\rangle$ are not eigenstates of $\hat{A}$, in which case

$$\begin{aligned}\langle\hat{A}\rangle &= \left[\langle 1|c_1^* + \langle 2|c_2^*\right]\hat{A}\left[c_1|1\rangle + c_2|2\rangle\right] \\ &= c_1^*c_1\langle 1|\hat{A}|1\rangle + c_1^*c_2\langle 1|\hat{A}|2\rangle + c_2^*c_1\langle 2|\hat{A}|1\rangle + c_2^*c_2\langle 2|\hat{A}|2\rangle.\end{aligned} \tag{7.14}$$

It is useful to consider $\langle r|\hat{A}|s\rangle$ as an element A_{rs} of a matrix $\boldsymbol{A}$:

As for vectors, the symbols for matrices are set in ***bold italic*** type.

$$\boldsymbol{A} = \begin{pmatrix} A_{11} & A_{12} \\ A_{21} & A_{22} \end{pmatrix} = \begin{pmatrix} \langle 1|\hat{A}|1\rangle & \langle 1|\hat{A}|2\rangle \\ \langle 2|\hat{A}|1\rangle & \langle 2|\hat{A}|2\rangle \end{pmatrix}. \tag{7.15}$$

This allows Eqn 7.14 to be written as

$$\langle\hat{A}\rangle = c_1^*c_1A_{11} + c_1^*c_2A_{12} + c_2^*c_1A_{21} + c_2^*c_2A_{22}, \tag{7.16}$$

or, much more compactly, using the vector notation introduced above:

$$\langle\hat{A}\rangle = \begin{pmatrix} c_1^* & c_2^* \end{pmatrix}\begin{pmatrix} A_{11} & A_{12} \\ A_{21} & A_{22} \end{pmatrix}\begin{pmatrix} c_1 \\ c_2 \end{pmatrix} = \boldsymbol{\psi}^\dagger \boldsymbol{A}\boldsymbol{\psi}. \tag{7.17}$$

Compare this expression with Eqns 7.8 and 7.10.

7.4 Angular momentum

We now turn from these general expressions to a more concrete example. One of the most important operators in the quantum mechanical description of any system is the *Hamiltonian*, which determines the total energy of the system,

$$E = \langle\hat{H}\rangle = \langle\psi|\hat{H}|\psi\rangle. \tag{7.18}$$

The *spin Hamiltonians* used in NMR look very different from more conventional *space Hamiltonians* of the form

$$\hat{H} = -\frac{\hbar^2}{2m}\nabla^2 + \hat{V}$$

with kinetic and potential energy operators.

As we shall see, the Hamiltonians which occur in NMR are closely related to the operators that describe angular momentum; these operators play a central role in the theory of NMR.

The energy of a classical magnetic dipole, such as a compass needle, in a magnetic field is given by $E = -\boldsymbol{\mu}\cdot\boldsymbol{B}_0 = -\mu_z B_z$, where $\boldsymbol{\mu}$ is the magnetic moment, μ_z is its z component, and the magnetic field, $\boldsymbol{B}_0$, is along the z-axis. For a particle such as an atomic nucleus, the magnetic moment is proportional to its angular momentum so that $\mu_z = \gamma I_z$, where the proportionality coefficient γ is the *magnetogyric ratio*. The analogous quantum mechanical

See Hore *Nuclear Magnetic Resonance* (2015) for a discussion of this expression.

expression can be obtained by replacing E and I_z by the corresponding quantum mechanical operators:

$$\hat{H} = -\gamma B_0 \hbar \hat{I}_z = \hbar\omega_0 \hat{I}_z \tag{7.19}$$

where $\omega_0 = -\gamma B_0$ is an angular frequency (the *Larmor* frequency) and $\hbar$ is Planck's constant divided by 2π. Note that we are following the most common convention in NMR, which defines the operator for the angular momentum around the z-axis as $\hbar\hat{I}_z$, not just $\hat{I}_z$. However, writing energies as multiples of $\hbar$ or, equivalently, working with units such that $\hbar = 1$, is a common and convenient practice in NMR, and so we can write

$$\hat{H} = \omega_0 \hat{I}_z. \tag{7.20}$$

Note that while Eqn 7.20 is true for any spin quantum number, these particular eigenstates are specific to spin-$\frac{1}{2}$ nuclei. Other notations are also used: in particular the states $|\alpha\rangle$ and $|\beta\rangle$ are often called $|+\frac{1}{2}\rangle$ and $|-\frac{1}{2}\rangle$ respectively. In this notation, Eqn 7.21 takes the simple form $\hat{I}_z|m\rangle = m|m\rangle$, where $\hbar m$ is the z component of the angular momentum.

The eigenstates of $\hat{I}_z$, and thus of the Hamiltonian, are conventionally called $|\alpha\rangle$ (spin-up) and $|\beta\rangle$ (spin-down); these states are a natural choice for basis states. The corresponding eigenvalues are given by

$$\hat{I}_z|\alpha\rangle = +\tfrac{1}{2}|\alpha\rangle \qquad \hat{I}_z|\beta\rangle = -\tfrac{1}{2}|\beta\rangle \tag{7.21}$$

and

$$\hat{H}|\alpha\rangle = +\tfrac{1}{2}\omega_0|\alpha\rangle \qquad \hat{H}|\beta\rangle = -\tfrac{1}{2}\omega_0|\beta\rangle. \tag{7.22}$$

Thus $|\alpha\rangle$ and $|\beta\rangle$ are separated in energy by $\hbar|\omega_0|$. For many nuclei, including ^{1}H, γ is positive and so ω_0 is negative; thus $|\alpha\rangle$ has a lower energy than $|\beta\rangle$.

Matrix representations are calculated as shown in Eqn 7.15. The matrix $\boldsymbol{I_z}$ is diagonal because $|\alpha\rangle$ and $|\beta\rangle$ are orthonormal eigenstates of $\hat{I}_z$.

The eigenvalues in Eqn 7.21 can be used to determine the *matrix representation* of $\hat{I}_z$:

$$\boldsymbol{I_z} = \begin{pmatrix} \frac{1}{2} & 0 \\ 0 & -\frac{1}{2} \end{pmatrix}. \tag{7.23}$$

From Eqn 7.13, the expectation value is

$$\langle \hat{I}_z \rangle = \tfrac{1}{2}|c_\alpha|^2 - \tfrac{1}{2}|c_\beta|^2, \tag{7.24}$$

i.e. half the difference between the probabilities of being found in the two states.

Justifying Eqns 7.21 and 7.25 is not a simple business. One approach is to note that these forms give the correct commutation relationships, as shown below (Eqn 7.30). See Susskind and Friedman *Quantum Mechanics: The Theoretical Minimum* (2014) for a more detailed discussion.

Unlike $\hat{I}_z$, the operators corresponding to x and y angular momentum do not have $|\alpha\rangle$ and $|\beta\rangle$ as eigenstates. Instead these operators interconvert $|\alpha\rangle$ and $|\beta\rangle$:

$$\begin{aligned} \hat{I}_x|\alpha\rangle &= \tfrac{1}{2}|\beta\rangle \qquad & \hat{I}_x|\beta\rangle &= \tfrac{1}{2}|\alpha\rangle \\ \hat{I}_y|\alpha\rangle &= \tfrac{1}{2}\mathrm{i}|\beta\rangle \qquad & \hat{I}_y|\beta\rangle &= -\tfrac{1}{2}\mathrm{i}|\alpha\rangle. \end{aligned} \tag{7.25}$$

Remember that $\mathrm{i} = \sqrt{-1}$. The matrices in Eqns 7.23 and 7.26 are closely related to the Pauli matrices. See Atkins and Friedman (2010) for a more detailed discussion.

As before, these expressions can be expressed more compactly in matrix form

$$\boldsymbol{I_x} = \begin{pmatrix} 0 & \frac{1}{2} \\ \frac{1}{2} & 0 \end{pmatrix} \qquad \boldsymbol{I_y} = \begin{pmatrix} 0 & -\frac{1}{2}\mathrm{i} \\ \frac{1}{2}\mathrm{i} & 0 \end{pmatrix} \tag{7.26}$$

From Eqn 7.16, their expectation values are

$$\begin{aligned}\langle \hat{I}_x \rangle &= \tfrac{1}{2}\left(c_\alpha c_\beta^* + c_\alpha^* c_\beta\right) = \mathrm{Re}\left(c_\alpha c_\beta^*\right)\\ \langle \hat{I}_y \rangle &= \tfrac{1}{2}\mathrm{i}\left(c_\alpha c_\beta^* - c_\alpha^* c_\beta\right) = -\mathrm{Im}\left(c_\alpha c_\beta^*\right).\end{aligned} \tag{7.27}$$

Re(a) and Im(a) indicate the real and imaginary parts of a.

Note that transverse magnetization depends on the *cross products* $c_\alpha c_\beta^*$ and $c_\alpha^* c_\beta$, while the longitudinal component z is determined by the *self-products* $c_\alpha c_\alpha^*$ and $c_\beta c_\beta^*$.

The cross products represent *coherent superposition states*, or more simply *coherences*. This is explored further in Chapter 8.

The matrix representations (Eqns 7.23 and 7.26) can be used to deduce further properties of the angular momentum operators:

$$\hat{I}_x^2 = \hat{I}_x \hat{I}_x = \tfrac{1}{4}\hat{1} \tag{7.28}$$

where $\hat{1}$ is the unit operator (and identical results for $\hat{I}_y$ and $\hat{I}_z$),

$$\hat{I}_x \hat{I}_y = -\hat{I}_y \hat{I}_x = \tfrac{1}{2}\mathrm{i}\hat{I}_z \tag{7.29}$$

$$\left[\hat{I}_x, \hat{I}_y\right] \equiv \hat{I}_x \hat{I}_y - \hat{I}_y \hat{I}_x = \mathrm{i}\hat{I}_z \tag{7.30}$$

(and equivalent expressions obtained by cyclic permutation of the subscripts, $x \to y \to z \to x$). The expression in Eqn 7.30, the *operator commutator*, will be of great importance later on.

These *commutation relations* must hold for any set of operators which describe angular momentum; effectively they can be taken as *defining* the forms of the operators. They can themselves be deduced from the general properties of angular momentum and the fundamental position-momentum commutation relationship, as described in Atkins and Friedman (2010).

7.5 **Free precession**

Now we have the tools needed to look at the NMR behaviour of a single isolated spin-$\tfrac{1}{2}$ particle. First, we consider the time dependence of the system during a period of free precession. This may be obtained from the *time-dependent Schrödinger equation*

$$\frac{\mathrm{d}}{\mathrm{d}t}|\psi\rangle = -\mathrm{i}\hat{H}|\psi\rangle \tag{7.31}$$

which has been written in units such that $\hbar = 1$. For the general initial state $|\psi\rangle = c_\alpha|\alpha\rangle + c_\beta|\beta\rangle$,

$$\frac{\mathrm{d}}{\mathrm{d}t}|\psi\rangle = \frac{\mathrm{d}c_\alpha}{\mathrm{d}t}|\alpha\rangle + \frac{\mathrm{d}c_\beta}{\mathrm{d}t}|\beta\rangle = -\mathrm{i}c_\alpha\hat{H}|\alpha\rangle - \mathrm{i}c_\beta\hat{H}|\beta\rangle. \tag{7.32}$$

Note that the basis states cannot vary with time, and so the time dependence of a state $|\psi\rangle$ must be due to variations in its coefficients.

Multiplying from the left by $\langle\alpha|$ gives

$$\frac{\mathrm{d}c_\alpha}{\mathrm{d}t}\langle\alpha|\alpha\rangle + \frac{\mathrm{d}c_\beta}{\mathrm{d}t}\langle\alpha|\beta\rangle = -\mathrm{i}c_\alpha\langle\alpha|\hat{H}|\alpha\rangle - \mathrm{i}c_\beta\langle\alpha|\hat{H}|\beta\rangle \tag{7.33}$$

or, using the fact that $|\alpha\rangle$ and $|\beta\rangle$ are orthonormal,

$$\frac{\mathrm{d}c_\alpha}{\mathrm{d}t} = -\mathrm{i}c_\alpha H_{\alpha\alpha} - \mathrm{i}c_\beta H_{\alpha\beta} \tag{7.34}$$

where the matrix $\boldsymbol{H}$ is defined using Eqn 7.15, as before. In the same way, multiplying Eqn 7.32 from the left by $\langle\beta|$ gives a corresponding expression for dc_β/dt. These expressions can be expressed more concisely in matrix form as

The dots indicate differentiation with respect to time.

$$\begin{pmatrix} \dot{c}_\alpha \\ \dot{c}_\beta \end{pmatrix} = -\mathrm{i} \begin{pmatrix} H_{\alpha\alpha} & H_{\alpha\beta} \\ H_{\beta\alpha} & H_{\beta\beta} \end{pmatrix} \begin{pmatrix} c_\alpha \\ c_\beta \end{pmatrix} \tag{7.35}$$

or

$$\dot{\boldsymbol{\psi}} = -\mathrm{i}\boldsymbol{H}\boldsymbol{\psi}. \tag{7.36}$$

The form of this equation is identical to that of the conventional time-dependent Schrödinger equation (Eqn 7.31), and could have been deduced directly using the analogy between operators and matrices (see Appendix C).

Using Eqns 7.20 and 7.23,

$$\boldsymbol{H} = \omega_0 \boldsymbol{I}_z = \begin{pmatrix} \frac{1}{2}\omega_0 & 0 \\ 0 & -\frac{1}{2}\omega_0 \end{pmatrix} \tag{7.37}$$

so that Eqn 7.35 gives

$$\dot{c}_\alpha = -\tfrac{1}{2}\mathrm{i}\omega_0 c_\alpha, \qquad \dot{c}_\beta = \tfrac{1}{2}\mathrm{i}\omega_0 c_\beta. \tag{7.38}$$

Integration of these expressions gives

$$c_\alpha(t) = c_\alpha(0)\mathrm{e}^{-\mathrm{i}\omega_0 t/2} \qquad c_\beta(t) = c_\beta(0)\mathrm{e}^{\mathrm{i}\omega_0 t/2} \tag{7.39}$$

i.e. the two coefficients oscillate at frequencies $\pm\omega_0/2$.

Suppose we start from the initial condition $c_\alpha(0) = c_\beta(0) = 1/\sqrt{2}$, i.e. $\langle \hat{I}_x \rangle = \frac{1}{2}$, $\langle \hat{I}_y \rangle = \langle \hat{I}_z \rangle = 0$; then from Eqns 7.24 and 7.27

$$\begin{aligned} \langle \hat{I}_x \rangle &= \mathrm{Re}\left[c_\alpha(0)c_\beta^*(0)\mathrm{e}^{-\mathrm{i}\omega_0 t} \right] = \tfrac{1}{2}\cos\omega_0 t \\ \langle \hat{I}_y \rangle &= -\mathrm{Im}\left[c_\alpha(0)c_\beta^*(0)\mathrm{e}^{-\mathrm{i}\omega_0 t} \right] = \tfrac{1}{2}\sin\omega_0 t \\ \langle \hat{I}_z \rangle &= \tfrac{1}{2}\left[c_\alpha(0)c_\alpha^*(0) - c_\beta(0)c_\beta^*(0) \right] = 0. \end{aligned} \tag{7.40}$$

This is *exactly* what is expected from the vector model for free precession in the laboratory frame (neglecting relaxation): the *z*-magnetization does not change, while the *x* and *y*-magnetizations oscillate at the Larmor frequency.

7.6 Radiofrequency pulses

Working in a rotating frame, as described in Appendix F, greatly simplifies things, and is almost essential for all but the simplest problems (see Section 1.3). Note the similarity between Eqns 7.41 and 7.20.

The effects of radiofrequency pulses can be treated in a similar way. At resonance, with the radiofrequency field B_1 taken along the *x*-axis of the rotating frame,

$$\hat{H} = -\gamma B_1 \hat{I}_x = \omega_1 \hat{I}_x. \tag{7.41}$$

Proceeding in the same manner as above gives

$$\dot{c}_\alpha = -\tfrac{1}{2}\mathrm{i}\omega_1 c_\beta \tag{7.42}$$

$$\dot{c}_\beta = -\tfrac{1}{2}\mathrm{i}\omega_1 c_\alpha. \tag{7.43}$$

These coupled differential equations can be uncoupled by differentiating Eqn 7.42 once again:

$$\ddot{c}_\alpha = \frac{\mathrm{d}}{\mathrm{d}t}\dot{c}_\alpha = -\tfrac{1}{2}\mathrm{i}\omega_1 \dot{c}_\beta = -\tfrac{1}{4}\omega_1^2 c_\alpha. \tag{7.44}$$

The general solution of this differential equation is $c_\alpha(t) = A\cos(\omega_1 t/2) + B\sin(\omega_1 t/2)$, where A and B are constants determined by the boundary conditions.

Starting from the spin-up state, $|\alpha\rangle$, i.e. $c_\alpha(0) = 1$ and $c_\beta(0) = 0$, Eqn 7.44 gives

$$c_\alpha(t) = \cos\left(\tfrac{1}{2}\omega_1 t\right). \tag{7.45}$$

Substituting this result back into Eqn 7.43 and integrating gives

$$c_\beta(t) = -\mathrm{i}\sin\left(\tfrac{1}{2}\omega_1 t\right). \tag{7.46}$$

Hence

$$\begin{aligned}
\langle \hat{I}_x \rangle &= \mathrm{Re}\left[\cos\left(\tfrac{1}{2}\omega_1 t\right)\mathrm{i}\sin\left(\tfrac{1}{2}\omega_1 t\right)\right] = 0 \\
\langle \hat{I}_y \rangle &= -\mathrm{Im}\left[\cos\left(\tfrac{1}{2}\omega_1 t\right)\mathrm{i}\sin\left(\tfrac{1}{2}\omega_1 t\right)\right] = -\tfrac{1}{2}\sin(\omega_1 t) \\
\langle \hat{I}_z \rangle &= \tfrac{1}{2}\left[\cos^2\left(\tfrac{1}{2}\omega_1 t\right) - \sin^2\left(\tfrac{1}{2}\omega_1 t\right)\right] = \tfrac{1}{2}\cos(\omega_1 t).
\end{aligned} \tag{7.47}$$

Thus, magnetization starting along the z-axis rotates at a frequency ω_1 around the x-axis towards the negative y-axis (cf. Sections 1.4 and 3.2), as expected.

7.7 Matrix and operator exponentials

The methods outlined above could be extended to deal with more complex NMR experiments: evolution under any Hamiltonian will lead to a set of (usually coupled) differential equations, which can then be solved. There is, however, a simpler approach. Rather than solving the problem for each individual Hamiltonian we find a general solution, which can then be evaluated for any particular case.

The matrix form of the time-dependent Schrödinger equation (Eqn 7.36) has the formal solution

This solution is only correct if $\boldsymbol{H}$ is constant in time, a point we come back to below.

$$\boldsymbol{\psi}(t) = \left[\boldsymbol{1} + (-\mathrm{i}\boldsymbol{H}t) + \frac{(-\mathrm{i}\boldsymbol{H}t)^2}{2!} + \frac{(-\mathrm{i}\boldsymbol{H}t)^3}{3!} + \cdots\right]\boldsymbol{\psi}(0) \tag{7.48}$$

which is easily confirmed by differentiating the right-hand side with respect to t. By analogy with the series expansion of the ordinary exponential function we can define the *matrix exponential* of a matrix $\boldsymbol{M}$ by

$$\mathrm{e}^{\boldsymbol{M}} = \boldsymbol{1} + \boldsymbol{M} + \frac{\boldsymbol{M}^2}{2!} + \frac{\boldsymbol{M}^3}{3!} + \cdots \tag{7.49}$$

More practical approaches for calculating matrix exponentials are described in Appendix D.

and so the general solution of the time-dependent Schrödinger equation in matrix form is

$$\boldsymbol{\psi}(t) = \mathrm{e}^{-\mathrm{i}\boldsymbol{H}t}\,\boldsymbol{\psi}(0). \tag{7.50}$$

As usual, an equivalent equation can be written for operators and kets:

$$|\psi(t)\rangle = \mathrm{e}^{-\mathrm{i}\hat{H}t}|\psi(0)\rangle. \tag{7.51}$$

General methods for evaluating matrix and operator exponentials are described in Appendix D, and the specific case of $\hat{I}_x$ is considered in Appendix E. Here we just quote the answer in two specific cases

$$\mathrm{e}^{-\mathrm{i}\omega_0 t \boldsymbol{I_z}} = \begin{pmatrix} \mathrm{e}^{-\mathrm{i}\omega_0 t/2} & 0 \\ 0 & \mathrm{e}^{\mathrm{i}\omega_0 t/2} \end{pmatrix} \tag{7.52}$$

$$\mathrm{e}^{-\mathrm{i}\omega_1 t \boldsymbol{I_x}} = \begin{pmatrix} \cos(\frac{1}{2}\omega_1 t) & -\mathrm{i}\sin(\frac{1}{2}\omega_1 t) \\ -\mathrm{i}\sin(\frac{1}{2}\omega_1 t) & \cos(\frac{1}{2}\omega_1 t) \end{pmatrix} \tag{7.53}$$

both of which can be verified by direct differentiation, using

$$\frac{\mathrm{d}}{\mathrm{d}t}\mathrm{e}^{-\mathrm{i}\boldsymbol{H}t} = -\mathrm{i}\boldsymbol{H}\mathrm{e}^{-\mathrm{i}\boldsymbol{H}t}. \tag{7.54}$$

As usual we are working in a rotating frame; this approach is justified in Appendix F.

These matrix exponentials can then be used to repeat the calculations of Sections 7.5 and 7.6. For the case of a radiofrequency pulse applied along the x-axis to the spin-up state, the solution is

$$\begin{pmatrix} c_\alpha(t) \\ c_\beta(t) \end{pmatrix} = \begin{pmatrix} \cos(\frac{1}{2}\omega_1 t) & -\mathrm{i}\sin(\frac{1}{2}\omega_1 t) \\ -\mathrm{i}\sin(\frac{1}{2}\omega_1 t) & \cos(\frac{1}{2}\omega_1 t) \end{pmatrix}\begin{pmatrix} 1 \\ 0 \end{pmatrix} = \begin{pmatrix} \cos(\frac{1}{2}\omega_1 t) \\ -\mathrm{i}\sin(\frac{1}{2}\omega_1 t) \end{pmatrix} \tag{7.55}$$

exactly as calculated before.

The approach above assumes that the Hamiltonian is constant in time, and so it might seem difficult to apply it to complicated NMR experiments where different radiofrequency pulses are applied at different times. The method can, however, be easily adapted to the case where the Hamiltonian takes some constant form $\boldsymbol{H}_1$ for some time t_1, and then some different form $\boldsymbol{H}_2$ for some time t_2, and so on. In this case the solution is

Hamiltonians of this kind are said to be *piecewise constant*.

$$\boldsymbol{\psi} = \mathrm{e}^{-\mathrm{i}\boldsymbol{H}_n t_n}\ldots\mathrm{e}^{-\mathrm{i}\boldsymbol{H}_2 t_2}\,\mathrm{e}^{-\mathrm{i}\boldsymbol{H}_1 t_1}\,\boldsymbol{\psi}(0) \tag{7.56}$$

obtained by evolving the initial state under each Hamiltonian in turn.

7.8 Summary

- NMR experiments can be described using conventional quantum mechanics.

- Angular momentum operators are required to describe the Hamiltonians of spin systems.
- A knowledge of vectors, matrices, and operators helps in describing the problem.
- A naïve approach requires the solution of coupled differential equations.
- A better approach is to write the general solution in terms of matrix or operator exponentials.

7.9 Exercises

Worked solutions to the exercises are available on the Online Resource Centre at www.oxfordtextbooks.co.uk/orc/hore2e/

7.1 Consider the kets $|\psi_a\rangle = |1\rangle + |2\rangle$, $|\psi_b\rangle = |1\rangle + \mathrm{i}|2\rangle$, and $|\psi_c\rangle = |1\rangle + \sqrt{3}\,\mathrm{i}|2\rangle$. Find the corresponding normalized kets $|\psi_a'\rangle$, and so on, and the corresponding normalized bras.

7.2 Calculate the inner products $\langle\psi_a|\psi_b\rangle$ and $\langle\psi_a'|\psi_b'\rangle$ and comment on your answers.

7.3 Calculate the inner products $\langle\psi_a'|\psi_c'\rangle$ and $\langle\psi_c'|\psi_a'\rangle$ and comment on your answers.

7.4 Write down the matrix forms of $\boldsymbol{\psi}_\alpha$ and $\boldsymbol{\psi}_\beta$ corresponding to the basis kets $|\alpha\rangle$ and $|\beta\rangle$. Hence show that the matrices $\boldsymbol{I_x}$, $\boldsymbol{I_y}$, and $\boldsymbol{I_z}$ (Eqns 7.23 and 7.26) act as described in Eqns 7.21 and 7.25.

7.5 Show that $[B,A] = -[A,B]$ and that $[A,B+C] = [A,B] + [A,C]$.

7.6 Determine the nine possible binary products of the matrices $\boldsymbol{I_x}$, $\boldsymbol{I_y}$, and $\boldsymbol{I_z}$, and hence show that these matrices have the correct commutation relations to represent angular momentum.

7.7 Use the methods in Appendices D and E to confirm Eqns 7.52 and 7.53. Find the corresponding result for $\boldsymbol{I_y}$.

7.8 Repeat the calculations in Section 7.5 using matrix exponentials. Repeat the calculations in Section 7.6 for a y-pulse.

8 Density matrices

8.1 Introduction

The introduction of operator exponentials provides a huge simplification in describing the evolution of NMR spin states, but the solution is still given as a vector of coefficients, which has to be further interpreted to obtain information about the observable NMR signal. In this chapter we go a step further, by introducing the density matrix description which allows the observable NMR signal to be immediately deduced. As we will see density matrices are essentially equivalent to product operators, and the rules for manipulating product operators are simple consequences of the evolution of density matrices under particular Hamiltonians.

Density matrices and density operators are described in many texts on quantum mechanics, and in all advanced treatments of NMR. A more detailed treatment can be found in Ernst, Bodenhausen, and Wokaun, *Principles of Nuclear Magnetic Resonance in One and Two Dimensions* (1985).

8.2 The density operator

As shown in the previous chapter, expectation values of an operator $\hat{A}$ depend on *products* of coefficients. For an operator acting on a single spin (Eqn 7.16)

$$\langle \hat{A} \rangle = c_\alpha^* c_\alpha A_{\alpha\alpha} + c_\alpha^* c_\beta A_{\alpha\beta} + c_\beta^* c_\alpha A_{\beta\alpha} + c_\beta^* c_\beta A_{\beta\beta}. \tag{8.1}$$

It is therefore useful to define a *density operator* which is most conveniently described in terms of the elements of its corresponding *density matrix*, $\boldsymbol{\rho}$:

The density operator can be derived in many different ways, with varying degrees of mathematical formality. Here we mostly use simple analogies.

$$\rho_{rs} = \langle r | \hat{\rho} | s \rangle = c_r c_s^*. \tag{8.2}$$

Combining Eqns 8.1 and 8.2,

$$\langle \hat{A} \rangle = \rho_{\alpha\alpha} A_{\alpha\alpha} + \rho_{\beta\alpha} A_{\alpha\beta} + \rho_{\alpha\beta} A_{\beta\alpha} + \rho_{\beta\beta} A_{\beta\beta} \tag{8.3}$$

which is simply the *trace* of the product of $\boldsymbol{\rho}$ and $\mathbf{A}$:

The trace of a matrix is the sum of the elements on its principal diagonal (see Appendix B).

$$\langle \hat{A} \rangle = \sum_j \sum_k \rho_{jk} A_{kj} = \sum_j (\boldsymbol{\rho} \mathbf{A})_{jj} = \mathrm{Tr}(\boldsymbol{\rho} \mathbf{A}). \tag{8.4}$$

Equation 8.2 shows how the density matrix $\boldsymbol{\rho}$ can be calculated directly from the underlying coefficients. For a single spin

$$\boldsymbol{\rho}=\begin{pmatrix}c_\alpha\\c_\beta\end{pmatrix}\begin{pmatrix}c_\alpha^* & c_\beta^*\end{pmatrix}=\begin{pmatrix}c_\alpha c_\alpha^* & c_\alpha c_\beta^*\\c_\beta c_\alpha^* & c_\beta c_\beta^*\end{pmatrix} \tag{8.5}$$

Note that the product of a column vector and a row vector is an $n \times n$ square *matrix*, while the product of a row vector and a column vector gives a single *number*.

or, in general,

$$\boldsymbol{\rho}=\boldsymbol{\psi}\boldsymbol{\psi}^{\dagger}. \tag{8.6}$$

Using the analogy between vectors and kets, the density operator can be written in terms of bras and kets:

$$\hat{\rho}=|\psi\rangle\langle\psi|. \tag{8.7}$$

In Chapter 7 we saw that the expectation value of $\hat{I}_z$ is determined by $c_\alpha c_\alpha^*$ and $c_\beta c_\beta^*$, while those of $\hat{I}_x$ and $\hat{I}_y$ depend on $c_\alpha c_\beta^*$ and $c_\beta c_\alpha^*$. Thus, the diagonal elements of the density matrix represent *populations*, while the off-diagonal elements represent *single-quantum coherences*.

See also Section 8.4. In general diagonal terms always represent populations, including ordered population states, while off-diagonal elements represent coherences of various orders.

Next we need to determine how $\boldsymbol{\rho}$ evolves under a given Hamiltonian. One way to approach this problem is to calculate the behaviour of a single element:

$$\dot{\rho}_{rs}=\dot{c}_r c_s^*+c_r\dot{c}_s^*. \tag{8.8}$$

This expression is obtained from Eqn 8.2 by the standard rules for differentiating a product. As in the previous chapter, the dot indicates differentiation with respect to time.

Generalizing from Eqn 7.35,

$$\begin{aligned}\dot{c}_r&=-\mathrm{i}\sum_k H_{rk}c_k\\ \dot{c}_s^*&=+\mathrm{i}\sum_k H_{sk}^*c_k^*\end{aligned} \tag{8.9}$$

and so

$$\begin{aligned}\dot{\rho}_{rs}&=-\mathrm{i}\sum_k H_{rk}c_k c_s^*+\mathrm{i}\sum_k c_r c_k^* H_{sk}^*\\&=-\mathrm{i}\sum_k H_{rk}\rho_{ks}+\mathrm{i}\sum_k \rho_{rk}H_{ks}\\&=-\mathrm{i}(\boldsymbol{H}\boldsymbol{\rho})_{rs}+\mathrm{i}(\boldsymbol{\rho}\boldsymbol{H})_{rs}=-\mathrm{i}[\boldsymbol{H},\rho]_{rs}\end{aligned} \tag{8.10}$$

We have used the fact that $\boldsymbol{H}$ is Hermitian (i.e. $\boldsymbol{H}=\boldsymbol{H}^{\dagger}$, or $H_{sk}^*=H_{ks}$); see Appendix B for elementary properties of matrices.

Since Eqn 8.10 holds for each of the elements of $\boldsymbol{\rho}$, it must hold for the entire matrix.

A more sophisticated approach is to begin from Eqn 8.7, and differentiate the operator product directly:

$$\frac{\mathrm{d}\hat{\rho}}{\mathrm{d}t}=\frac{\mathrm{d}}{\mathrm{d}t}|\psi\rangle\langle\psi|=\left(\frac{\mathrm{d}}{\mathrm{d}t}|\psi\rangle\right)\langle\psi|+|\psi\rangle\left(\frac{\mathrm{d}}{\mathrm{d}t}\langle\psi|\right). \tag{8.11}$$

Since $\langle\psi|=(|\psi\rangle)^{\dagger}$, Eqn 7.31 gives

$$\frac{\mathrm{d}}{\mathrm{d}t}\langle\psi|=\left(-\mathrm{i}\hat{H}|\psi\rangle\right)^{\dagger}=\mathrm{i}\langle\psi|\hat{H} \tag{8.12}$$

Recall that $(\hat{A}\hat{B})^{\dagger}=\hat{B}^{\dagger}\hat{A}^{\dagger}$; see Appendices B and C.

and so

$$\frac{\mathrm{d}\hat{\rho}}{\mathrm{d}t}=-\mathrm{i}\hat{H}\hat{\rho}+\mathrm{i}\hat{\rho}\hat{H}=-\mathrm{i}\left[\hat{H},\hat{\rho}\right]. \qquad (8.13)$$

This equation, which relates the evolution of the density operator to the Hamiltonian, plays a central role in the theory of density matrices. It is important enough to have a name: the *Liouville-von Neumann equation*.

8.3 Solving the Liouville-von Neumann equation

Approaches for calculating matrix and operator exponentials are described in Appendices D and E.

The Liouville-von Neumann equation is readily solved when the Hamiltonian is constant for some time, t. The solution is most easily written in terms of *operator exponentials*:

$$\hat{\rho}(t)=\mathrm{e}^{-\mathrm{i}\hat{H}t}\hat{\rho}(0)\mathrm{e}^{+\mathrm{i}\hat{H}t}. \qquad (8.14)$$

The initial density operator is, of course, independent of t.

This solution can be verified by differentiating Eqn 8.14:

$$\begin{aligned}\frac{\mathrm{d}\hat{\rho}}{\mathrm{d}t}&=\left(\frac{\mathrm{d}\mathrm{e}^{-\mathrm{i}\hat{H}t}}{\mathrm{d}t}\right)\hat{\rho}(0)\mathrm{e}^{+\mathrm{i}\hat{H}t}+\mathrm{e}^{-\mathrm{i}\hat{H}t}\hat{\rho}(0)\left(\frac{\mathrm{d}\mathrm{e}^{+\mathrm{i}\hat{H}t}}{\mathrm{d}t}\right)\\&=-\mathrm{i}\hat{H}\mathrm{e}^{-\mathrm{i}\hat{H}t}\hat{\rho}(0)\mathrm{e}^{+\mathrm{i}\hat{H}t}+\mathrm{e}^{-\mathrm{i}\hat{H}t}\hat{\rho}(0)\mathrm{e}^{+\mathrm{i}\hat{H}t}\,\mathrm{i}\hat{H}\\&=-\mathrm{i}\hat{H}\hat{\rho}+\mathrm{i}\hat{\rho}\hat{H}=-\mathrm{i}\left[\hat{H},\hat{\rho}\right],\end{aligned} \qquad (8.15)$$

Note that any operator commutes with any power of itself, and therefore with its operator exponential as well.

where we have used the fact that

$$\frac{\mathrm{d}}{\mathrm{d}t}\mathrm{e}^{k\hat{A}t}=k\hat{A}\,\mathrm{e}^{k\hat{A}t}=\mathrm{e}^{k\hat{A}t}k\hat{A} \qquad (8.16)$$

for any real or complex number k.

For a discussion of this, see Appendix D.

The Hamiltonian operator $\hat{H}$ is always Hermitian, and this has the consequence that the corresponding operator exponential is unitary. We introduce the *unitary propagator*

$$\hat{U}=\mathrm{e}^{-\mathrm{i}\hat{H}t}, \qquad (8.17)$$

and its adjoint

$$\hat{U}^{\dagger}=\mathrm{e}^{+\mathrm{i}\hat{H}t}, \qquad (8.18)$$

which allow Eqn 8.14 to be written more simply as

$$\hat{\rho}(t)=\hat{U}\hat{\rho}(0)\hat{U}^{\dagger}, \qquad (8.19)$$

where $\hat{\rho}(0)$ is the initial density operator.

Solving the Liouville-von Neumann equation is more difficult when the Hamiltonian varies with time. If, however, the Hamiltonian is constant during individual periods making up the entire evolution time, then the equation is easily solved by an extension of Eqn 8.14:

$$\hat{\rho}(0) \xrightarrow{\hat{H}_1 t_1} e^{-i\hat{H}_1 t_1}\hat{\rho}(0)e^{+i\hat{H}_1 t_1} \xrightarrow{\hat{H}_2 t_2} e^{-i\hat{H}_2 t_2}e^{-i\hat{H}_1 t_1}\hat{\rho}(0)e^{+i\hat{H}_1 t_1}e^{+i\hat{H}_2 t_2} \tag{8.20}$$

Compare this expression with Eqn 7.56.

and so on. Equivalently,

$$\hat{\rho}(0) \xrightarrow{\hat{H}_1 t_1} \hat{U}_1\hat{\rho}(0)\hat{U}_1^{\dagger} \xrightarrow{\hat{H}_2 t_2} \hat{U}_2\hat{U}_1\hat{\rho}(0)\hat{U}_1^{\dagger}\hat{U}_2^{\dagger}. \tag{8.21}$$

8.4 Ensemble averages

The density operator is certainly a convenient mathematical device, which avoids explicitly calculating coefficients, such as c_α, when it is only their products, such as $c_\alpha c_\beta^*$, which actually determine the values of observable quantities (e.g. Eqns 7.24 and 7.27). It has a second great advantage, however, as it provides an elegant way of treating macroscopic systems.

Many macroscopic systems, such as NMR samples, are made up of a very large number of identical independent copies (effectively an ensemble of copies) of individual microscopic systems. Each microscopic system feels the same basic interactions, and thus evolves under the same Hamiltonian, but the systems do not all evolve in the same way because each copy has its own initial state.

By an ensemble of independent copies we mean that the individual microscopic systems do not interact with one another in any way.

Consider, for example, two microscopic systems with initial states $|\psi_1\rangle$ and $|\psi_2\rangle$, respectively. Using the bra-ket approach outlined in the previous chapter it is necessary to evaluate the behaviour of each system individually: it is not possible simply to add the two kets together. The same problem occurs with density operators, but in this case a great simplification can be made as long as it is not necessary to determine individual expectation values for the two microscopic systems, but only their sum. This is a realistic restriction, as it corresponds exactly to performing experiments on a macroscopic system containing two or more indistinguishable microscopic systems. In this case,

$$\langle\hat{A}\rangle_1 + \langle\hat{A}\rangle_2 = \mathrm{Tr}(\boldsymbol{\rho}_1\boldsymbol{A}) + \mathrm{Tr}(\boldsymbol{\rho}_2\boldsymbol{A}) = \mathrm{Tr}([\boldsymbol{\rho}_1 + \boldsymbol{\rho}_2]\boldsymbol{A}), \tag{8.22}$$

so it is possible to add density operators before evaluating expectation values. Similarly,

Fundamentally this approach works because all the operations involved are *linear*; see Appendix B.

$$\begin{aligned}\hat{\rho}_1(t) + \hat{\rho}_2(t) &= e^{-i\hat{H}t}\hat{\rho}_1(0)e^{+i\hat{H}t} + e^{-i\hat{H}t}\hat{\rho}_2(0)e^{+i\hat{H}t} \\ &= e^{-i\hat{H}t}\left[\hat{\rho}_1(0) + \hat{\rho}_2(0)\right]e^{+i\hat{H}t}.\end{aligned} \tag{8.23}$$

Thus it is possible to define a summed density operator

$$\hat{\rho}(t) = \hat{\rho}_1(t) + \hat{\rho}_2(t) \tag{8.24}$$

which can be used in any further calculations.

The same approach can be used with any number of microscopic systems: the corresponding density operators can be summed to produce an overall density operator. However, it is more usual to *average* the individual density operators. Thus, in general, the density operator of a macroscopic system is simply the ensemble average of the density operators of the underlying microscopic systems,

Remember that this approach is only applicable when the microscopic systems are *independent* and have *identical spin Hamiltonians*. If this is not the case it is necessary to use a larger density matrix which explicitly describes each microscopic system.

$$\hat{\rho} = \sum_i p_i \hat{\rho}_i, \tag{8.25}$$

where p_i is the probability of a microscopic system being in state $\hat{\rho}_i$ and, like all probabilities, the p_i are positive and sum to 1. The individual elements of $\boldsymbol{\rho}$ are given by

$$\rho_{rs} = \overline{c_r c_s^*} \tag{8.26}$$

where the bar indicates the ensemble average in Eqn 8.25.

These ensemble averaged density matrices describe *mixed states*, rather than the *pure states* used in simple descriptions of quantum mechanics. A particularly important example of an ensemble-averaged density operator is that of a macroscopic system at thermal equilibrium, which is the usual starting point for an NMR experiment.

Pure states have little relevance in conventional NMR, although they are produced in some exotic experiments. It can be useful, however, to use pure state notation to describe spin Hamiltonians in some complex pulse sequences. Appendix G shows how to write pure states in product operator notation.

To see the difference between pure and mixed states consider two different ensembles. In both, the microscopic systems are described by

$$|\psi\rangle = a_\alpha e^{i\phi_\alpha}|\alpha\rangle + a_\beta e^{i\phi_\beta}|\beta\rangle, \tag{8.27}$$

where a_α and a_β are amplitudes, and ϕ_α and ϕ_β are phases, and all four quantities are real. The corresponding density matrix is

Note that a_α and a_β are the magnitudes of the complex coefficients in Eqn 7.1.

$$\boldsymbol{\rho} = \begin{pmatrix} a_\alpha^2 & a_\alpha a_\beta e^{i(\phi_\alpha - \phi_\beta)} \\ a_\alpha a_\beta e^{-i(\phi_\alpha - \phi_\beta)} & a_\beta^2 \end{pmatrix}. \tag{8.28}$$

In the first ensemble (a *pure* state) all the systems are in the same state, and so the density matrix of the whole ensemble is identical to that of each of the microscopic systems. The second ensemble (a *mixed* state) is similar to the first, except that the microscopic systems are in subtly different states, with the phase parameters ϕ_α and ϕ_β varying randomly from one system to the next. The macroscopic density matrix is obtained by averaging over the ensemble, i.e. over all values of ϕ_α and ϕ_β. The (random) phase factors $\exp\left[\pm i(\phi_\alpha - \phi_\beta)\right]$ interfere destructively to give

$$\boldsymbol{\rho} = \begin{pmatrix} a_\alpha^2 & 0 \\ 0 & a_\beta^2 \end{pmatrix}. \tag{8.29}$$

Comparing these two examples, we see that the diagonal elements of the density matrices, which correspond to the populations of the two basis states, are the same in both cases, but the *off-diagonal* elements are quite different. These off-diagonal elements, which arise from systems in superposition states, are only non-zero when there is *phase coherence* between the states of the microscopic systems. Such *phase coherent superpositions* are called *coherences*. In NMR, they are created by applying one or more pulses of coherent radiofrequency radiation. These coherences are then dephased by T_2 relaxation.

Equation 8.29 can be rewritten in a slightly simpler form. Since a_α^2 and a_β^2 are positive numbers, and $a_\alpha^2 + a_\beta^2 = 1$, they can be considered as probabilities, and so

$$\boldsymbol{\rho} = \begin{pmatrix} p_\alpha & 0 \\ 0 & p_\beta \end{pmatrix} \tag{8.30}$$

or

$$\hat{\rho} = p_\alpha |\alpha\rangle\langle\alpha| + p_\beta |\beta\rangle\langle\beta| \tag{8.31}$$

Compare this expression with Eqns 8.7 and 8.25.

which is a mixture of spins in the states $|\alpha\rangle$ and $|\beta\rangle$. In a thermal equilibrium state, the probabilities of spins being in the two states, p_α and p_β, are given by the Boltzmann distribution, and the off-diagonal elements of the density matrix have been averaged away, as discussed above.

The thermal equilibrium density operator can be rewritten:

$$\begin{aligned}\boldsymbol{\rho} &= \tfrac{1}{2}\begin{pmatrix} p_\alpha + p_\beta & 0 \\ 0 & p_\alpha + p_\beta \end{pmatrix} + \tfrac{1}{2}\begin{pmatrix} p_\alpha - p_\beta & 0 \\ 0 & -p_\alpha + p_\beta \end{pmatrix} \\ &= \tfrac{1}{2}\mathbf{1} + (p_\alpha - p_\beta)\mathbf{I_z}.\end{aligned} \tag{8.32}$$

The unit matrix, $\mathbf{1}$, commutes with everything, and so by Eqn 8.15 it does not evolve under any Hamiltonian. Furthermore, it cannot be observed in any NMR experiment, which follows from Eqn 8.4 and the fact that all NMR observables have zero trace. Thus all the interesting behaviour may be deduced by examining the evolution of the $\mathbf{I_z}$ term. Finally, $(p_\alpha - p_\beta)$ is just a scaling factor and is often omitted, e.g. in Eqn 8.35 below.

8.5 Application to NMR

To illustrate how density matrices can be used, we now calculate the effect of an on-resonance x-pulse in the rotating frame, with the system initially in the mixed state $\hat{I}_z$. As before, the Hamiltonian is given by

Note that we use the same notation to describe both the state of the system and our operations upon it.

$$\hat{H} = \omega_1 \hat{I}_x \tag{8.33}$$

so that (using Eqn 8.14)

$$\boldsymbol{\rho}(t) = \mathrm{e}^{-\mathrm{i}\omega_1 t \mathbf{I_x}} \boldsymbol{\rho}(0) \mathrm{e}^{+\mathrm{i}\omega_1 t \mathbf{I_x}} \tag{8.34}$$

with

$$\boldsymbol{\rho}(0) = \boldsymbol{I_z}. \tag{8.35}$$

Details of this calculation are given in Appendix E.

To make further progress we need explicit expressions for the exponential matrices in Eqn 8.34. As discussed in Section 7.7,

$$e^{\pm i\omega_1 t \boldsymbol{I_x}} = \begin{pmatrix} \cos(\frac{1}{2}\omega_1 t) & \pm i\sin(\frac{1}{2}\omega_1 t) \\ \pm i\sin(\frac{1}{2}\omega_1 t) & \cos(\frac{1}{2}\omega_1 t) \end{pmatrix}. \tag{8.36}$$

We can then evaluate $\boldsymbol{\rho}(t)$ by multiplying matrices according to Eqn 8.34 to obtain

$$\boldsymbol{\rho}(t) = \begin{pmatrix} \frac{1}{2}\cos\omega_1 t & \frac{1}{2}i\sin\omega_1 t \\ -\frac{1}{2}i\sin\omega_1 t & -\frac{1}{2}\cos\omega_1 t \end{pmatrix}. \tag{8.37}$$

Finally we can use Eqn 8.4 to calculate expectation values:

$$\begin{aligned} \langle \hat{I}_x \rangle &= \mathrm{Tr}(\boldsymbol{\rho I_x}) = \mathrm{Tr}\begin{pmatrix} \frac{1}{4}i\sin\omega_1 t & \frac{1}{4}\cos\omega_1 t \\ -\frac{1}{4}\cos\omega_1 t & -\frac{1}{4}i\sin\omega_1 t \end{pmatrix} = 0 \\ \langle \hat{I}_y \rangle &= \mathrm{Tr}(\boldsymbol{\rho I_y}) = \mathrm{Tr}\begin{pmatrix} -\frac{1}{4}\sin\omega_1 t & -\frac{1}{4}i\cos\omega_1 t \\ -\frac{1}{4}i\cos\omega_1 t & -\frac{1}{4}\sin\omega_1 t \end{pmatrix} = -\frac{1}{2}\sin\omega_1 t \\ \langle \hat{I}_z \rangle &= \mathrm{Tr}(\boldsymbol{\rho I_z}) = \mathrm{Tr}\begin{pmatrix} \frac{1}{4}\cos\omega_1 t & -\frac{1}{4}i\sin\omega_1 t \\ -\frac{1}{4}i\sin\omega_1 t & \frac{1}{4}\cos\omega_1 t \end{pmatrix} = \frac{1}{2}\cos\omega_1 t. \end{aligned} \tag{8.38}$$

cf. Fig. 1.2.

As expected from the vector model, a radiofrequency pulse along the *x*-axis causes the expectation values of $\hat{I}_y$ and $\hat{I}_z$ to oscillate, at frequency ω_1, while the expectation value of $\hat{I}_x$ remains constant at zero.

This calculation can be extended to include the effects of *off-resonance* pulses. In this case the Hamiltonian is

$$\boldsymbol{H} = \omega_1 \boldsymbol{I_x} + \Omega \boldsymbol{I_z} = \begin{pmatrix} \frac{1}{2}\Omega & \frac{1}{2}\omega_1 \\ \frac{1}{2}\omega_1 & -\frac{1}{2}\Omega \end{pmatrix}, \tag{8.39}$$

where Ω is the offset frequency in the rotating frame. This can be rewritten in terms of the *off-resonance fraction*,

$$f = \Omega/\omega_1, \tag{8.40}$$

to get

$$\boldsymbol{H} = \omega_1(\boldsymbol{I_x} + f\boldsymbol{I_z}) = \frac{1}{2}\omega_1 \begin{pmatrix} f & 1 \\ 1 & -f \end{pmatrix}. \tag{8.41}$$

Evaluating the corresponding propagator by hand is now quite complicated, but the calculation is easy with assistance from a computer, and the result is

Some suitable computer programs are discussed in Appendix D.

$$e^{-i\boldsymbol{H}t} = \begin{pmatrix} \cos\left[\tfrac{1}{2}r\omega_1 t\right] - \dfrac{if}{r}\sin\left[\tfrac{1}{2}r\omega_1 t\right] & -\dfrac{i}{r}\sin\left[\tfrac{1}{2}r\omega_1 t\right] \\ -\dfrac{i}{r}\sin\left[\tfrac{1}{2}r\omega_1 t\right] & \cos\left[\tfrac{1}{2}r\omega_1 t\right] + \dfrac{if}{r}\sin\left[\tfrac{1}{2}r\omega_1 t\right] \end{pmatrix} \tag{8.42}$$

with

$$r = \sqrt{1+f^2}\,. \tag{8.43}$$

Note that when f is small, so that $r \approx 1$, the result is very similar to the on-resonance propagator (Eqn 8.36), while when f is large so that $r \approx f$, the off-diagonal terms in the propagator become small. Thus when the offset frequency is comparable to or larger than the nutation frequency the pulse does not effectively excite the spin. Thus pulses with relatively small nutation frequencies will be *frequency selective*, as discussed in Appendix H.

8.6 Connection to product operators

The approach just outlined is ideal for numerical calculations on a computer, especially for more complex problems where analytical expressions for matrix exponentials are hard or impossible to obtain. In some cases, however, there is a more subtle and convenient method which exploits the properties of the operators involved rather than relying on somewhat tiresome matrix arithmetic. It also gives insight into the link between product operators and density matrices.

Consider a system described by an initial density operator at $t = 0$,

As usual we use operators (e.g. $\hat{A}$) and their matrix representations (**A**) interchangeably.

$$\boldsymbol{\rho}(0) = \boldsymbol{A} \tag{8.44}$$

(e.g. $\boldsymbol{A} = \boldsymbol{I_z}$, as in Section 8.5), and imagine that we wish to determine the evolution of a spin system under the influence of a Hamiltonian

$$\boldsymbol{H} = b\boldsymbol{B} \tag{8.45}$$

(e.g. $b = \omega_1$, $\boldsymbol{B} = \boldsymbol{I_x}$, as in Eqn 8.33). Suppose that the following commutation relations hold:

$$\begin{aligned} [\boldsymbol{A},\boldsymbol{B}] &= i\boldsymbol{C}, \\ [\boldsymbol{B},\boldsymbol{C}] &= i\boldsymbol{A}. \end{aligned} \tag{8.46}$$

Under these admittedly special circumstances, the evolution of the density operator is particularly simple:

$$\boldsymbol{\rho}(t) = \boldsymbol{A}\cos bt - \boldsymbol{C}\sin bt. \tag{8.47}$$

This important relation can be justified as follows. With the definition of $\boldsymbol{H}$ given above, the Liouville-von Neumann equation is

$$\frac{\mathrm{d}\boldsymbol{\rho}}{\mathrm{d}t} = -\mathrm{i}b[\boldsymbol{B},\boldsymbol{\rho}]. \tag{8.48}$$

Remember that $[\boldsymbol{B},\boldsymbol{A}] = -[\boldsymbol{A},\boldsymbol{B}]$.

Substituting Eqn 8.47 into the right-hand side gives:

$$\begin{aligned} -\mathrm{i}b[\boldsymbol{B},\boldsymbol{\rho}] &= -\mathrm{i}b[\boldsymbol{B},(\boldsymbol{A}\cos bt - \boldsymbol{C}\sin bt)] \\ &= -\mathrm{i}b[\boldsymbol{B},\boldsymbol{A}]\cos bt + \mathrm{i}b[\boldsymbol{B},\boldsymbol{C}]\sin bt \\ &= -\mathrm{i}b(-\mathrm{i}\boldsymbol{C})\cos bt + \mathrm{i}b(\mathrm{i}\boldsymbol{A})\sin bt \\ &= -b\boldsymbol{C}\cos bt - b\boldsymbol{A}\sin bt, \end{aligned} \tag{8.49}$$

which is precisely the same as the left-hand side of Eqn 8.48:

$$\frac{\mathrm{d}\boldsymbol{\rho}}{\mathrm{d}t} = \frac{\mathrm{d}}{\mathrm{d}t}(\boldsymbol{A}\cos bt - \boldsymbol{C}\sin bt) = -b\boldsymbol{A}\sin bt - b\boldsymbol{C}\cos bt. \tag{8.50}$$

As a check that Eqn 8.47 is applicable, the relevant commutators are:
$[\boldsymbol{A},\boldsymbol{B}] = [\boldsymbol{I_z},\boldsymbol{I_x}] = \mathrm{i}\boldsymbol{I_y} = \mathrm{i}\boldsymbol{C}$;
$[\boldsymbol{B},\boldsymbol{C}] = [\boldsymbol{I_x},\boldsymbol{I_y}] = \mathrm{i}\boldsymbol{I_z} = \mathrm{i}\boldsymbol{A}$.

To see how Eqn 8.47 works, we first repeat the calculation at the start of Section 8.5 (evolution during an on-resonance radiofrequency pulse). The three operators involved are $\boldsymbol{A} = \boldsymbol{I_z}$, $\boldsymbol{B} = \boldsymbol{I_x}$, and $\boldsymbol{C} = \boldsymbol{I_y}$, with $b = \omega_1$. Thus,

$$\begin{aligned} \boldsymbol{\rho}(t) &= \boldsymbol{I_z}\cos\omega_1 t - \boldsymbol{I_y}\sin\omega_1 t \\ &= \begin{pmatrix} \frac{1}{2} & 0 \\ 0 & -\frac{1}{2} \end{pmatrix}\cos\omega_1 t - \begin{pmatrix} 0 & -\frac{1}{2}\mathrm{i} \\ \frac{1}{2}\mathrm{i} & 0 \end{pmatrix}\sin\omega_1 t \\ &= \begin{pmatrix} \frac{1}{2}\cos\omega_1 t & \frac{1}{2}\mathrm{i}\sin\omega_1 t \\ -\frac{1}{2}\mathrm{i}\sin\omega_1 t & -\frac{1}{2}\cos\omega_1 t \end{pmatrix} \end{aligned} \tag{8.51}$$

as in Eqn 8.37.

Now consider the free precession that occurs after a 90°_x pulse: $\boldsymbol{A} = -\boldsymbol{I_y}$, $\boldsymbol{B} = \boldsymbol{I_z}$, $\boldsymbol{C} = -\boldsymbol{I_x}$, and $b = \Omega$ so that

The commutators are:
$[\boldsymbol{A},\boldsymbol{B}] = [-\boldsymbol{I_y},\boldsymbol{I_z}] = -\mathrm{i}\boldsymbol{I_x} = \mathrm{i}\boldsymbol{C}$;
$[\boldsymbol{B},\boldsymbol{C}] = [\boldsymbol{I_z},-\boldsymbol{I_x}] = -\mathrm{i}\boldsymbol{I_y} = \mathrm{i}\boldsymbol{A}$.

$$\boldsymbol{\rho}(t) = -\boldsymbol{I_y}\cos\Omega t + \boldsymbol{I_x}\sin\Omega t \tag{8.52}$$

again as expected (see Figs. 1.4 and 3.2).

Expectation values can be obtained directly from equations such as Eqn 8.52, e.g.

Note that $\mathrm{Tr}[\boldsymbol{I_j}\boldsymbol{I_k}] = \frac{1}{2}$ if $j = k$, or 0 if $j \neq k$, where $j,k = x,y,z$.

$$\begin{aligned} \langle \hat{I}_y \rangle &= \mathrm{Tr}[\boldsymbol{\rho}\boldsymbol{I_y}] = -\mathrm{Tr}[\boldsymbol{I_y^2}]\cos\Omega t + \mathrm{Tr}[\boldsymbol{I_x}\boldsymbol{I_y}]\sin\Omega t \\ &= -\mathrm{Tr}\begin{pmatrix} \frac{1}{4} & 0 \\ 0 & \frac{1}{4} \end{pmatrix}\cos\Omega t + \mathrm{Tr}\begin{pmatrix} \frac{1}{4}\mathrm{i} & 0 \\ 0 & -\frac{1}{4}\mathrm{i} \end{pmatrix}\sin\Omega t = -\tfrac{1}{2}\cos\Omega t, \end{aligned} \tag{8.53}$$

and similarly $\langle \hat{I}_x \rangle = \frac{1}{2}\sin\Omega t$. This approach can also be applied in the general case. For a one-spin density matrix, with the contribution from the identity operator removed (as usual), the corresponding product operator representation would be simply:

The correspondence of product operators and density matrices, and the use of Eqns 8.44 and 8.51, will be pursued further in Chapter 9. For two-spin density matrices, introduced there, the sum runs over all 15 operators.
By convention, product operators do not have 'hats', i.e. I_x not $\hat{I}_x$.

$$I_x\langle \hat{I}_x \rangle + I_y\langle \hat{I}_y \rangle + I_z\langle \hat{I}_z \rangle = I_x\mathrm{Tr}[\hat{\rho}\hat{I}_x] + I_y\mathrm{Tr}[\hat{\rho}\hat{I}_y] + I_z\mathrm{Tr}[\hat{\rho}\hat{I}_z] \tag{8.54}$$

where, as usual, quantum mechanical operators have hats and product operators are bare-headed.

All of this should be looking pretty familiar from Chapters 3, 4, and 5 of Part A. If we replace the matrices by the corresponding product operators, Eqn 8.52 is *exactly* the result of the product operator formalism. To emphasize the connection between the two approaches, we can rewrite Eqn 8.47 as:

$$\mathbf{A} \xrightarrow{bt\mathbf{B}} \mathbf{A}\cos bt - \mathbf{C}\sin bt. \tag{8.55}$$

Thus, product operators are nothing more than a cunning way of doing analytical density matrix calculations.

8.7 Summary

- The density matrix provides an alternative description of a spin state, which makes it easy to calculate the observable signal.
- Density matrices are also essential to describe mixed states, such as spin systems at thermal equilibrium.
- The description of mixed states can be simplified by considering only the portion which evolves under Hamiltonians.
- The evolution can be calculated using matrix exponentials, but it is often simpler to use analytical short cuts.
- Product operators are in fact nothing more than a cunning way of doing analytical density matrix calculations.

8.8 Exercises

Worked solutions to the exercises are available on the Online Resource Centre at www.oxfordtextbooks.co.uk/orc/hore2e/

8.1 Consider the ket $|\psi_a\rangle = \cos(\theta/2)|\alpha\rangle + \sin(\theta/2)e^{i\phi}|\beta\rangle$. Show that $|\psi_a\rangle$ is normalized and find the corresponding density matrix $\boldsymbol{\rho}_a$.

8.2 Show that $\boldsymbol{\rho}_a$ is Hermitian and has trace 1.

8.3 Consider the ket $|\boldsymbol{\psi}_b\rangle = e^{i\gamma}|\boldsymbol{\psi}_a\rangle$. Find $\boldsymbol{\rho}_b$ and comment on your answer.

8.4 Show that mixed state density matrices are Hermitian and have trace 1. (Hint: consider Eqn 8.25.)

8.5 Does the matrix $\mathbf{I_z}$ represent a mixed state? If not, does it matter?

8.6 Confirm the results of Eqns 8.37 and 8.38.

8.7 Repeat these calculations to find the evolution of $\mathbf{I_z}$ under $\mathbf{H} = \omega_1 \mathbf{I_y}$.

8.8 Use similar calculations to calculate the result of the spin echo shown in Fig. 1.6 applied to an isolated nucleus with offset frequency Ω. Determine the density matrix at each point in the spin echo sequence and interpret it as a linear combination of $\mathbf{I_x}$, $\mathbf{I_y}$, and $\mathbf{I_z}$.

8.9 Repeat this calculation using the methods in Section 8.6.

Weak coupling

9.1 Introduction

We have now seen all the machinery needed to use density matrices to predict the behaviour of isolated nuclei. As in the case of product operators, the density matrix approach reassuringly reproduces the results of the vector model. With two or more coupled spins, however, density matrices start to come into their own. Here we discuss the case of weakly coupled spins in order to see more clearly the correspondence between product operators and density matrices and to derive some of the results we quoted without proof in Part A.

Weak coupling occurs when the J-coupling constant is small compared with the difference between the offset frequencies of the spins, i.e. when $|2\pi J_{\mathrm{IS}}| \ll |\Omega_{\mathrm{I}} - \Omega_{\mathrm{S}}|$.

9.2 Density operators in two-spin systems

The methods outlined in the previous two chapters can easily be extended to a two-spin (IS) system, which will have four basis vectors:

$$|\psi\rangle = c_{\alpha\alpha}|\alpha_{\mathrm{I}}\alpha_{\mathrm{S}}\rangle + c_{\alpha\beta}|\alpha_{\mathrm{I}}\beta_{\mathrm{S}}\rangle + c_{\beta\alpha}|\beta_{\mathrm{I}}\alpha_{\mathrm{S}}\rangle + c_{\beta\beta}|\beta_{\mathrm{I}}\beta_{\mathrm{S}}\rangle. \quad (9.1)$$

Matrix representations of the six one-spin angular momentum operators, $\boldsymbol{I_x}$, $\boldsymbol{I_y}$, $\boldsymbol{I_z}$, $\boldsymbol{S_x}$, $\boldsymbol{S_y}$, and $\boldsymbol{S_z}$, can be obtained using Eqns 7.21 and 7.25, e.g.

$$\hat{I}_x|\alpha_{\mathrm{I}}\alpha_{\mathrm{S}}\rangle = \tfrac{1}{2}|\beta_{\mathrm{I}}\alpha_{\mathrm{S}}\rangle, \qquad \hat{S}_y|\beta_{\mathrm{I}}\alpha_{\mathrm{S}}\rangle = \tfrac{1}{2}\mathrm{i}|\beta_{\mathrm{I}}\beta_{\mathrm{S}}\rangle \quad (9.2)$$

(note that I operators do not affect spin S and vice versa). Thus, for example,

$$\boldsymbol{I_x} = \begin{pmatrix} 0 & 0 & \frac{1}{2} & 0 \\ 0 & 0 & 0 & \frac{1}{2} \\ \frac{1}{2} & 0 & 0 & 0 \\ 0 & \frac{1}{2} & 0 & 0 \end{pmatrix}. \quad (9.3)$$

There is, however, a simple approach which allows matrix representations to be deduced directly. Two-spin matrices can be obtained by taking *direct products* of one-spin matrices (Eqns 7.23 and 7.26) and the unit matrix. Thus,

Direct products are explained in Appendix I.

$$I_y = \begin{pmatrix} 0 & -\frac{1}{2}i \\ \frac{1}{2}i & 0 \end{pmatrix} \otimes \begin{pmatrix} 1 & 0 \\ 0 & 1 \end{pmatrix} = \begin{pmatrix} 0 & 0 & -\frac{1}{2}i & 0 \\ 0 & 0 & 0 & -\frac{1}{2}i \\ \frac{1}{2}i & 0 & 0 & 0 \\ 0 & \frac{1}{2}i & 0 & 0 \end{pmatrix}, \qquad (9.4)$$

and so on. Matrices corresponding to spin S are calculated in just the same way, except that the order of the two matrices in the direct product is reversed:

$$S_y = \begin{pmatrix} 1 & 0 \\ 0 & 1 \end{pmatrix} \otimes \begin{pmatrix} 0 & -\frac{1}{2}i \\ \frac{1}{2}i & 0 \end{pmatrix} = \begin{pmatrix} 0 & -\frac{1}{2}i & 0 & 0 \\ \frac{1}{2}i & 0 & 0 & 0 \\ 0 & 0 & 0 & -\frac{1}{2}i \\ 0 & 0 & \frac{1}{2}i & 0 \end{pmatrix}. \qquad (9.5)$$

Note that, as expected, all I matrices commute with all S matrices, e.g. $[I_x, S_y] = 0$. Moreover, the commutation and other relations given in Eqns 7.28–7.30 still hold, e.g. $I_x^2 = 1/4$, $I_x I_y = iI_z/2$, and $[I_x, I_y] = iI_z$.

Matrix representations of the other two-spin operators can be obtained in much the same way, either by taking direct products of the corresponding one-spin matrices,

$$2I_x S_y = 2\times \begin{pmatrix} 0 & \frac{1}{2} \\ \frac{1}{2} & 0 \end{pmatrix} \otimes \begin{pmatrix} 0 & -\frac{1}{2}i \\ \frac{1}{2}i & 0 \end{pmatrix} = \begin{pmatrix} 0 & 0 & 0 & -\frac{1}{2}i \\ 0 & 0 & \frac{1}{2}i & 0 \\ 0 & -\frac{1}{2}i & 0 & 0 \\ \frac{1}{2}i & 0 & 0 & 0 \end{pmatrix}, \qquad (9.6)$$

or by multiplying the corresponding two-spin matrices,

$$2I_y S_x = 2\times \begin{pmatrix} 0 & 0 & -\frac{1}{2}i & 0 \\ 0 & 0 & 0 & -\frac{1}{2}i \\ \frac{1}{2}i & 0 & 0 & 0 \\ 0 & \frac{1}{2}i & 0 & 0 \end{pmatrix} \begin{pmatrix} 0 & \frac{1}{2} & 0 & 0 \\ \frac{1}{2} & 0 & 0 & 0 \\ 0 & 0 & 0 & \frac{1}{2} \\ 0 & 0 & \frac{1}{2} & 0 \end{pmatrix}$$

$$= \begin{pmatrix} 0 & 0 & 0 & -\frac{1}{2}i \\ 0 & 0 & -\frac{1}{2}i & 0 \\ 0 & \frac{1}{2}i & 0 & 0 \\ \frac{1}{2}i & 0 & 0 & 0 \end{pmatrix}. \qquad (9.7)$$

For convenience we give a complete table of the two-spin matrices in Appendix I.

It was noted in Chapter 8 that the off-diagonal elements of the density matrix represent the *coherences* we have previously categorized as zero, single, double, etc. according to the difference in magnetic quantum number m of the two states involved. This may be seen more clearly from the matrix representations of the two-spin operators. For example, the off-diagonal elements of $\boldsymbol{I_x}$ (Eqn 9.3) connect pairs of states such as $|\alpha_I\alpha_S\rangle$ and $|\beta_I\alpha_S\rangle$, for which $\Delta m = \pm 1$, i.e. single-quantum coherences. Operator products such as $2\boldsymbol{I_x S_y}$ in Eqn 9.6 have matrix elements in the off-diagonal corners which connect $|\alpha_I\alpha_S\rangle$ and $|\beta_I\beta_S\rangle$ ($\Delta m = \pm 2$, double-quantum coherence) as well as elements linking $|\alpha_I\beta_S\rangle$ and $|\beta_I\alpha_S\rangle$ ($\Delta m = 0$, zero-quantum coherence).

Eqn 4.5 gives the product operator forms of zero- and double-quantum coherence.

9.3 *J*-coupling

The Hamiltonian describing the J-coupling between spins I and S has the general form

$$2\pi J \hat{I}\cdot\hat{S} \tag{9.8}$$

Henceforth, the symbol J_{IS} is abbreviated to J. The coupling Hamiltonian takes this form because it depends only on the relative orientation of the two spins: recall that the dot product of two vectors depends on the cosine of the angle between them (see Appendix B). See Hore (2015) for a more detailed discussion.

where J, the spin–spin coupling constant, is the strength of the coupling (coupling constants are traditionally measured in Hz, and so it is necessary to multiply by 2π to convert them into angular frequency units). The scalar product can be expanded as the sum of three terms

$$\hat{I}\cdot\hat{S} = \hat{I}_x\hat{S}_x + \hat{I}_y\hat{S}_y + \hat{I}_z\hat{S}_z \tag{9.9}$$

so that the matrix form of the coupling Hamiltonian is

$$2\pi J\left(\boldsymbol{I_x S_x} + \boldsymbol{I_y S_y} + \boldsymbol{I_z S_z}\right) = 2\pi J\begin{pmatrix} \frac{1}{4} & 0 & 0 & 0 \\ 0 & -\frac{1}{4} & \frac{1}{2} & 0 \\ 0 & \frac{1}{2} & -\frac{1}{4} & 0 \\ 0 & 0 & 0 & \frac{1}{4} \end{pmatrix}. \tag{9.10}$$

Evolution under the scalar coupling does not occur in isolation; one must consider the entire Hamiltonian, including the (Zeeman) interactions with the magnetic field. For a system of two coupled spins the complete Hamiltonian in the rotating frame has the form

$$\hat{H}_{IS} = \Omega_I\hat{I}_z + \Omega_S\hat{S}_z + 2\pi J\hat{I}\cdot\hat{S} \tag{9.11}$$

or

$$\boldsymbol{H_{IS}} = \begin{pmatrix} \frac{1}{2}(\Omega_I + \Omega_S) + \frac{1}{2}\pi J & 0 & 0 & 0 \\ 0 & \frac{1}{2}(\Omega_I - \Omega_S) - \frac{1}{2}\pi J & \pi J & 0 \\ 0 & \pi J & -\frac{1}{2}(\Omega_I - \Omega_S) - \frac{1}{2}\pi J & 0 \\ 0 & 0 & 0 & -\frac{1}{2}(\Omega_I + \Omega_S) + \frac{1}{2}\pi J \end{pmatrix} \tag{9.12}$$

In order to determine the evolution produced by $\boldsymbol{H}_{IS}$ it is necessary to calculate its matrix exponential, as described in Chapter 10. Before doing this, however, it is useful to consider an approximate approach.

9.4 Weak coupling: a brute-force approach

The Hamiltonian matrix described by Eqn 9.12 is almost diagonal. The weak coupling approximation simply neglects the two off-diagonal elements, and uses the simpler form

$$\boldsymbol{H}'_{IS} = \begin{pmatrix} \frac{1}{2}(\Omega_I+\Omega_S)+\frac{1}{2}\pi J & 0 & 0 & 0 \\ 0 & \frac{1}{2}(\Omega_I-\Omega_S)-\frac{1}{2}\pi J & 0 & 0 \\ 0 & 0 & -\frac{1}{2}(\Omega_I-\Omega_S)-\frac{1}{2}\pi J & 0 \\ 0 & 0 & 0 & -\frac{1}{2}(\Omega_I+\Omega_S)+\frac{1}{2}\pi J \end{pmatrix} \quad (9.13)$$

which is equivalent to

$$\boldsymbol{H}'_{IS} = \Omega_I \boldsymbol{I_z} + \Omega_S \boldsymbol{S_z} + 2\pi J \boldsymbol{I_z S_z}. \quad (9.14)$$

As discussed below, neglecting off-diagonal terms is equivalent to assuming that only the eigenvalues are changed by the coupling, while the eigenvectors remain unaffected. This corresponds to first-order perturbation theory, as described in Atkins and Friedman (2010). Comparison of Eqns 9.11 and 9.14 shows that it is the $\boldsymbol{I_x S_x}$ and $\boldsymbol{I_y S_y}$ terms that give rise to the discarded off-diagonal elements.

As will be seen later, this approximation is valid when $|2\pi J| \ll |\Omega_I - \Omega_S|$. With this simplification, it is clear that the eigenstates of the Hamiltonian are simply the basis states, $|\alpha_I\alpha_S\rangle$, $|\alpha_I\beta_S\rangle$, $|\beta_I\alpha_S\rangle$, and $|\beta_I\beta_S\rangle$, and that the eigenvalues are the diagonal elements:

$$\begin{aligned} \lambda_1 &= +\tfrac{1}{2}(\Omega_I+\Omega_S)+\tfrac{1}{2}\pi J \qquad & \lambda_2 &= +\tfrac{1}{2}(\Omega_I-\Omega_S)-\tfrac{1}{2}\pi J \\ \lambda_3 &= -\tfrac{1}{2}(\Omega_I-\Omega_S)-\tfrac{1}{2}\pi J \qquad & \lambda_4 &= -\tfrac{1}{2}(\Omega_I+\Omega_S)+\tfrac{1}{2}\pi J. \end{aligned} \quad (9.15)$$

To calculate the free induction decay expected for a weakly coupled pair of homonuclear spins, we first consider the effect of a 90°_x pulse on the equilibrium state $\rho(0) = \boldsymbol{I_z} + \boldsymbol{S_z}$. The Hamiltonian during the pulse is (Appendix F):

$$\boldsymbol{H} = \omega_1(\boldsymbol{I_x} + \boldsymbol{S_x}). \quad (9.16)$$

Its exponential is easily obtained because $\boldsymbol{I_x}$ and $\boldsymbol{S_x}$ commute, so that

$$e^{\pm i\boldsymbol{H}t} = e^{\pm i\omega_1 t(\boldsymbol{I_x}+\boldsymbol{S_x})} = e^{\pm i\omega_1 t\boldsymbol{I_x}} e^{\pm i\omega_1 t\boldsymbol{S_x}}. \quad (9.17)$$

N.B. $e^{\boldsymbol{A}+\boldsymbol{B}} = e^{\boldsymbol{A}} e^{\boldsymbol{B}} = e^{\boldsymbol{B}} e^{\boldsymbol{A}}$ if $[\boldsymbol{A},\boldsymbol{B}]=0$, as discussed in Appendix K.

Hence, using the single-spin 2×2 exponential matrix in Eqn 8.28, we have:

$$e^{\pm i\boldsymbol{H}t} = \begin{pmatrix} c & \pm is \\ \pm is & c \end{pmatrix} \otimes \begin{pmatrix} c & \pm is \\ \pm is & c \end{pmatrix} = \begin{pmatrix} c^2 & \pm isc & \pm isc & -s^2 \\ \pm isc & c^2 & -s^2 & \pm isc \\ \pm isc & -s^2 & c^2 & \pm isc \\ -s^2 & \pm isc & \pm isc & c^2 \end{pmatrix}, \quad (9.18)$$

with

$$c = \cos(\tfrac{1}{2}\omega_1 t), \qquad s = \sin(\tfrac{1}{2}\omega_1 t). \tag{9.19}$$

Matrix multiplication, using $\omega_1 t = \pi/2$ (i.e. $c = s = 1/\sqrt{2}$), gives

$$\boldsymbol{\rho}(t) = \mathrm{e}^{-\mathrm{i}\boldsymbol{H}t}\,\boldsymbol{\rho}(0)\mathrm{e}^{+\mathrm{i}\boldsymbol{H}t} = \begin{pmatrix} 0 & +\frac{1}{2}\mathrm{i} & +\frac{1}{2}\mathrm{i} & 0 \\ -\frac{1}{2}\mathrm{i} & 0 & 0 & +\frac{1}{2}\mathrm{i} \\ -\frac{1}{2}\mathrm{i} & 0 & 0 & +\frac{1}{2}\mathrm{i} \\ 0 & -\frac{1}{2}\mathrm{i} & -\frac{1}{2}\mathrm{i} & 0 \end{pmatrix} = -(\boldsymbol{I_y} + \boldsymbol{S_y}). \tag{9.20}$$

This is only an approximation, and an exact calculation could be performed as described in Section 8.5.

Thus, a 90°_x pulse rotates the z-magnetization of both spins onto the $-y$-axis. Neither the resonance offsets nor the J-coupling affects the excitation because of the assumption implicit in Eqn 9.16 that the radiofrequency field is the dominant interaction in the rotating frame.

The evolution during a subsequent period of free precession is easily calculated because the Hamiltonian is diagonal. Thus, in terms of the eigenvalues of Eqn 9.15:

$$\mathrm{e}^{\pm\mathrm{i}\boldsymbol{H}'_{IS}t} = \begin{pmatrix} \mathrm{e}^{\pm\mathrm{i}\lambda_1 t} & 0 & 0 & 0 \\ 0 & \mathrm{e}^{\pm\mathrm{i}\lambda_2 t} & 0 & 0 \\ 0 & 0 & \mathrm{e}^{\pm\mathrm{i}\lambda_3 t} & 0 \\ 0 & 0 & 0 & \mathrm{e}^{\pm\mathrm{i}\lambda_4 t} \end{pmatrix} \tag{9.21}$$

and

Here $\omega_{12} = \lambda_1 - \lambda_2$, and so on, are the differences between the eigenvalues for two eigenstates connected by an observable NMR transition.

$$\begin{aligned}\boldsymbol{\rho}(t) &= \mathrm{e}^{-\mathrm{i}\boldsymbol{H}'_{IS}t}(-\boldsymbol{I_y} - \boldsymbol{S_y})\mathrm{e}^{+\mathrm{i}\boldsymbol{H}'_{IS}t} \\ &= \begin{pmatrix} 0 & \frac{1}{2}\mathrm{i}\mathrm{e}^{-\mathrm{i}(\lambda_1-\lambda_2)t} & \frac{1}{2}\mathrm{i}\mathrm{e}^{-\mathrm{i}(\lambda_1-\lambda_3)t} & 0 \\ -\frac{1}{2}\mathrm{i}\mathrm{e}^{\mathrm{i}(\lambda_1-\lambda_2)t} & 0 & 0 & \frac{1}{2}\mathrm{i}\mathrm{e}^{-\mathrm{i}(\lambda_2-\lambda_4)t} \\ -\frac{1}{2}\mathrm{i}\mathrm{e}^{\mathrm{i}(\lambda_1-\lambda_3)t} & 0 & 0 & \frac{1}{2}\mathrm{i}\mathrm{e}^{-\mathrm{i}(\lambda_3-\lambda_4)t} \\ 0 & -\frac{1}{2}\mathrm{i}\mathrm{e}^{\mathrm{i}(\lambda_2-\lambda_4)t} & -\frac{1}{2}\mathrm{i}\mathrm{e}^{\mathrm{i}(\lambda_3-\lambda_4)t} & 0 \end{pmatrix} \\ &= \begin{pmatrix} 0 & \frac{1}{2}\mathrm{i}\mathrm{e}^{-\mathrm{i}\omega_{12}t} & \frac{1}{2}\mathrm{i}\mathrm{e}^{-\mathrm{i}\omega_{13}t} & 0 \\ -\frac{1}{2}\mathrm{i}\mathrm{e}^{\mathrm{i}\omega_{12}t} & 0 & 0 & \frac{1}{2}\mathrm{i}\mathrm{e}^{-\mathrm{i}\omega_{24}t} \\ -\frac{1}{2}\mathrm{i}\mathrm{e}^{\mathrm{i}\omega_{13}t} & 0 & 0 & \frac{1}{2}\mathrm{i}\mathrm{e}^{-\mathrm{i}\omega_{34}t} \\ 0 & -\frac{1}{2}\mathrm{i}\mathrm{e}^{\mathrm{i}\omega_{24}t} & -\frac{1}{2}\mathrm{i}\mathrm{e}^{\mathrm{i}\omega_{34}t} & 0 \end{pmatrix}.\end{aligned} \tag{9.22}$$

The expectation values of the observable in-phase x and y-magnetizations for spin I are therefore:

$$\begin{aligned}
\langle \boldsymbol{I_y} \rangle &= \mathrm{Tr}\left[\boldsymbol{\rho}(t)\boldsymbol{I_y}\right] \\
&= -\tfrac{1}{2}\cos[\omega_{13}t] - \tfrac{1}{2}\cos[\omega_{24}t] \\
&= -\tfrac{1}{2}\cos[(\Omega_I + \pi J)t] - \tfrac{1}{2}\cos[(\Omega_I - \pi J)t] \\
&= -\cos\Omega_I t \cos\pi Jt \\
\langle \boldsymbol{I_x} \rangle &= \mathrm{Tr}\left[\boldsymbol{\rho}(t)\boldsymbol{I_x}\right] \\
&= +\tfrac{1}{2}\sin[\omega_{13}t] + \tfrac{1}{2}\sin[\omega_{24}t] \\
&= +\tfrac{1}{2}\sin[(\Omega_I + \pi J)t] + \tfrac{1}{2}\sin[(\Omega_I - \pi J)t] \\
&= +\sin\Omega_I t \cos\pi Jt
\end{aligned} \qquad (9.23)$$

and for the I-spin antiphase operators:

$$\begin{aligned}
\langle 2\boldsymbol{I_y S_z} \rangle &= 2\mathrm{Tr}\left[\boldsymbol{\rho}(t)\boldsymbol{I_y S_z}\right] \\
&= -\tfrac{1}{2}\cos[\omega_{13}t] + \tfrac{1}{2}\cos[\omega_{24}t] \\
&= -\tfrac{1}{2}\cos[(\Omega_I + \pi J)t] + \tfrac{1}{2}\cos[(\Omega_I - \pi J)t] \\
&= +\sin\Omega_I t \sin\pi Jt \\
\langle 2\boldsymbol{I_x S_z} \rangle &= 2\mathrm{Tr}\left[\boldsymbol{\rho}(t)\boldsymbol{I_x S_z}\right] \\
&= +\tfrac{1}{2}\sin[\omega_{13}t] - \tfrac{1}{2}\sin[\omega_{24}t] \\
&= +\tfrac{1}{2}\sin[(\Omega_I + \pi J)t] - \tfrac{1}{2}\sin[(\Omega_I - \pi J)t] \\
&= +\cos\Omega_I t \sin\pi Jt
\end{aligned} \qquad (9.24)$$

in agreement with Eqn 3.12. The frequencies $\Omega_I \pm \pi J$ correspond to the two components of the I-spin doublet. The signs in front of the $\cos[(\Omega_I \pm \pi J)t]$ and $\sin[(\Omega_I \pm \pi J)t]$ terms indicate in-phase or antiphase doublets. Similar expressions can, of course, be obtained for the S spins.

9.5 Weak coupling: a more cunning approach

Alternatively, and with *much less effort*, we can use the approach introduced in Section 8.6 and summarized in Eqn 8.55:

Remember that operators and their matrix representations are used interchangeably.

$$\begin{aligned}
&\boldsymbol{A} \xrightarrow{bt\boldsymbol{B}} \boldsymbol{A}\cos bt - \boldsymbol{C}\sin bt \\
&\qquad [\boldsymbol{A},\boldsymbol{B}] = \mathrm{i}\boldsymbol{C} \\
&\qquad [\boldsymbol{B},\boldsymbol{C}] = \mathrm{i}\boldsymbol{A}.
\end{aligned} \qquad (9.25)$$

We first note that the three parts of the Hamiltonian (Eqn 9.14) all commute with each other. This allows us to calculate their effects sequentially and in any order.

This should be recognized as one of the properties of product operators that make them so convenient; see Appendix K.

As a check that Eqn 8.34 is applicable, the relevant commutators are:

$[\mathbf{A},\mathbf{B}] = -\mathrm{i}(2\mathbf{I_x S_z} + 2\mathbf{I_z S_x}) = \mathrm{i}\mathbf{C};$

$[\mathbf{B},\mathbf{C}] = -\mathrm{i}(\mathbf{I_y} + \mathbf{S_y}) = \mathrm{i}\mathbf{A}.$

During the period of free precession following the 90° pulse, the evolution under the J-coupling is described by the operators: $\mathbf{A} = -(\mathbf{I_y} + \mathbf{S_y})$, $\mathbf{B} = 2\mathbf{I_z S_z}$, $\mathbf{C} = -(2\mathbf{I_x S_z} + 2\mathbf{I_z S_x})$, $b = \pi J$, so that

$$-(\mathbf{I_y} + \mathbf{S_y}) \longrightarrow -(\mathbf{I_y} + \mathbf{S_y})\cos \pi Jt + (2\mathbf{I_x S_z} + 2\mathbf{I_z S_x})\sin \pi Jt \tag{9.26}$$

thus explaining the origin of the antiphase product operators, e.g. $2I_xS_z$, the factor of two in front of them, and the fact that the evolution under the J-coupling occurs at frequency πJ (Eqn 3.8). As in the case of isolated spins in Chapter 8, we see the correspondence of product operators and density matrices.

One can then repeat the process for each of the Zeeman interactions in turn, just as one would do with product operators, to obtain the results in Eqns 9.23 and 9.24.

9.6 Spin echoes

A spin echo occurs when two periods of free precession under the spin Hamiltonian, both of length τ, are separated by a 180° pulse. In Part A we showed that the effect of a *homonuclear* spin echo, in which the 180° pulse is applied to both spin I and spin S, is that the spins evolve only under the J-coupling, with the resonance offsets of the two spins being refocused. However, we only demonstrated that this was true for particular initial states, and merely asserted that it was always true. Here we use propagators to show that spin echoes do indeed work in the general case.

The evolution under a spin echo pulse sequence is given by the generalization of Eqn 8.21

$$\begin{aligned}\hat{\rho}(0) &\xrightarrow{\hat{H}_1 t_1} \hat{U}_1\hat{\rho}(0)\hat{U}_1^\dagger \xrightarrow{\hat{H}_2 t_2} \hat{U}_2\hat{U}_1\hat{\rho}(0)\hat{U}_1^\dagger\hat{U}_2^\dagger,\\ &\xrightarrow{\hat{H}_3 t_3} \hat{U}_3\hat{U}_2\hat{U}_1\hat{\rho}(0)\hat{U}_1^\dagger\hat{U}_2^\dagger\hat{U}_3^\dagger\end{aligned} \tag{9.27}$$

where the propagators are given by

$$\hat{U}_j = \mathrm{e}^{-\mathrm{i}\hat{H}_j t_j}. \tag{9.28}$$

So far we have thought of these propagators as being applied sequentially to the initial state, but they can instead be combined into a single *sequence propagator*

This works because matrix multiplication is associative; see Appendix B.

$$\begin{aligned}&\hat{\rho}(0) \rightarrow \hat{U}\hat{\rho}(0)\hat{U}^\dagger\\ &\hat{U} = \hat{U}_3\hat{U}_2\hat{U}_1.\end{aligned} \tag{9.29}$$

For the spin echo $\mathbf{H}_1$ and $\mathbf{H}_3$ are given by Eqn 9.14, with $t_1 = t_3 = \tau$, while $\mathbf{H}_2$ is given by Eqn 9.16 with $\omega_1 t_2 = \pi$ (we assume a 180°_x pulse for simplicity). Thus,

In Part A we considered a 180°_y pulse, but the difference is not important.

$$\mathbf{U} = \mathrm{e}^{-\mathrm{i}\tau(\Omega_I \mathbf{I_z} + \Omega_S \mathbf{S_z} + \pi J 2\mathbf{I_z S_z})}\,\mathrm{e}^{-\mathrm{i}\pi(\mathbf{I_x} + \mathbf{S_x})}\,\mathrm{e}^{-\mathrm{i}\tau(\Omega_I \mathbf{I_z} + \Omega_S \mathbf{S_z} + \pi J 2\mathbf{I_z S_z})}. \tag{9.30}$$

As the three terms in the background Hamiltonian all commute they can be calculated separately in any order, and the same is true of the pulses on the two spins. Thus,

$$\mathbf{U} = \mathrm{e}^{-\mathrm{i}\tau\pi J 2\mathbf{I_z S_z}}\,\mathrm{e}^{-\mathrm{i}\tau\Omega_I \mathbf{I_z}}\,\mathrm{e}^{-\mathrm{i}\tau\Omega_S \mathbf{S_z}}\,\mathrm{e}^{-\mathrm{i}\pi \mathbf{I_x}}\,\mathrm{e}^{-\mathrm{i}\pi \mathbf{S_x}}\,\mathrm{e}^{-\mathrm{i}\tau\Omega_S \mathbf{S_z}}\,\mathrm{e}^{-\mathrm{i}\tau\Omega_I \mathbf{I_z}}\,\mathrm{e}^{-\mathrm{i}\tau\pi J 2\mathbf{I_z S_z}}. \tag{9.31}$$

Since terms only involving spin I commute with terms only involving spin S, this can be reordered:

$$\boldsymbol{U} = \mathrm{e}^{-\mathrm{i}\tau\pi J 2\boldsymbol{I_z S_z}} \left(\mathrm{e}^{-\mathrm{i}\tau\Omega_I \boldsymbol{I_z}}\, \mathrm{e}^{-\mathrm{i}\pi \boldsymbol{I_x}}\, \mathrm{e}^{-\mathrm{i}\tau\Omega_I \boldsymbol{I_z}}\right)\left(\mathrm{e}^{-\mathrm{i}\tau\Omega_S \boldsymbol{S_z}}\, \mathrm{e}^{-\mathrm{i}\pi \boldsymbol{S_x}}\, \mathrm{e}^{-\mathrm{i}\tau\Omega_S \boldsymbol{S_z}}\right) \mathrm{e}^{-\mathrm{i}\tau\pi J 2\boldsymbol{I_z S_z}}. \quad (9.32)$$

Next the bracketed terms can be evaluated; for example, using matrix methods

This calculation can also be done using product operator methods of course. Note that since all the terms only involve spin I it is possible to perform this calculation using single-spin 2×2 matrices.

$$\begin{aligned}
\mathrm{e}^{-\mathrm{i}\tau\Omega_I \boldsymbol{I_z}}\mathrm{e}^{-\mathrm{i}\pi \boldsymbol{I_x}}\mathrm{e}^{-\mathrm{i}\tau\Omega_I \boldsymbol{I_z}} &= \begin{pmatrix} \mathrm{e}^{-\mathrm{i}\tau\Omega_I/2} & 0 \\ 0 & \mathrm{e}^{\mathrm{i}\tau\Omega_I/2} \end{pmatrix}\begin{pmatrix} 0 & -\mathrm{i} \\ -\mathrm{i} & 0 \end{pmatrix}\begin{pmatrix} \mathrm{e}^{-\mathrm{i}\tau\Omega_I/2} & 0 \\ 0 & \mathrm{e}^{\mathrm{i}\tau\Omega_I/2} \end{pmatrix} \\
&= \begin{pmatrix} 0 & -\mathrm{i} \\ -\mathrm{i} & 0 \end{pmatrix} \\
&= \mathrm{e}^{-\mathrm{i}\pi \boldsymbol{I_x}} \qquad (9.33)
\end{aligned}$$

and similarly for spin S. Thus,

$$\boldsymbol{U} = \mathrm{e}^{-\mathrm{i}\tau\pi J 2\boldsymbol{I_z S_z}}\, \mathrm{e}^{-\mathrm{i}\pi \boldsymbol{I_x}}\, \mathrm{e}^{-\mathrm{i}\pi \boldsymbol{S_x}}\, \mathrm{e}^{-\mathrm{i}\tau\pi J 2\boldsymbol{I_z S_z}} \qquad (9.34)$$

and the frequency offset terms have disappeared from the propagator. The evolution under the J-coupling can be calculated in the same way, but we now need to use 4×4 matrices. The combined propagator for the two 180°_x pulses is

$$\begin{aligned}
\mathrm{e}^{-\mathrm{i}\pi \boldsymbol{I_x}}\mathrm{e}^{-\mathrm{i}\pi \boldsymbol{S_x}} &= \begin{pmatrix} 0 & 0 & -\mathrm{i} & 0 \\ 0 & 0 & 0 & -\mathrm{i} \\ -\mathrm{i} & 0 & 0 & 0 \\ 0 & -\mathrm{i} & 0 & 0 \end{pmatrix}\begin{pmatrix} 0 & -\mathrm{i} & 0 & 0 \\ -\mathrm{i} & 0 & 0 & 0 \\ 0 & 0 & 0 & -\mathrm{i} \\ 0 & 0 & -\mathrm{i} & 0 \end{pmatrix} \\
&= \begin{pmatrix} 0 & 0 & 0 & -1 \\ 0 & 0 & -1 & 0 \\ 0 & -1 & 0 & 0 \\ -1 & 0 & 0 & 0 \end{pmatrix}, \qquad (9.35)
\end{aligned}$$

which *commutes* with the coupling Hamiltonian, and therefore commutes with the coupling propagators. This makes it easy to calculate the final result, which is

In this case the two propagators commute even though the Hamiltonians do not.

$$\begin{aligned}
\boldsymbol{U} &= \mathrm{e}^{-\mathrm{i}2\tau\pi J 2\boldsymbol{I_z S_z}}\, \mathrm{e}^{-\mathrm{i}\pi(\boldsymbol{I_x}+\boldsymbol{S_x})} \\
&= \mathrm{e}^{-\mathrm{i}\pi(\boldsymbol{I_x}+\boldsymbol{S_x})}\, \mathrm{e}^{-\mathrm{i}2\tau\pi J 2\boldsymbol{I_z S_z}}. \qquad (9.36)
\end{aligned}$$

Thus the spin echo can be replaced by the 180°_x pulse and free evolution under just the J-coupling for the total time 2τ.

An even simpler situation arises when two 180° pulses are used. Consider the case of free evolution for a total time 4τ divided into three parts by 180°_x pulses applied at times τ and 3τ. This is identical to two spin echoes applied back to back, and so the total propagator is

$$\boldsymbol{U} = \mathrm{e}^{-\mathrm{i}2\tau\pi J 2\boldsymbol{I_z S_z}}\, \mathrm{e}^{-\mathrm{i}\pi(\boldsymbol{I_x}+\boldsymbol{S_x})}\, \mathrm{e}^{-\mathrm{i}\pi(\boldsymbol{I_x}+\boldsymbol{S_x})}\, \mathrm{e}^{-\mathrm{i}2\tau\pi J 2\boldsymbol{I_z S_z}}. \qquad (9.37)$$

Here we have used the two different forms for the spin echo propagator in Eqn 9.36 in order to bring the two 180°_x pulses together. These two pulses can

The propagator for a 360°_x pulse on a single spin is in fact $-\mathbf{1}$, not $\mathbf{1}$, an effect known as *spinor behaviour*. However, in a two-spin system, as seen here, the two minus signs cancel out.

The concept of an average Hamiltonian can be applied in much more complex situations; see Ernst, Bodenhausen, and Wokaun (1985) for a more detailed treatment.

be combined to form a single 360°_x pulse, with the propagator $\mathbf{1}$, so the overall propagator is

$$\begin{aligned} \mathbf{U} &= e^{-i2\tau\pi J2\mathbf{I}_z\mathbf{S}_z}\,\mathbf{1}\,e^{-i2\tau\pi J2\mathbf{I}_z\mathbf{S}_z} \\ &= e^{-i4\tau\pi J2\mathbf{I}_z\mathbf{S}_z}. \end{aligned} \tag{9.38}$$

This is *identical* to evolution under just the J-coupling for the total time 4τ. Thus, one can say that the *average Hamiltonian* during the double spin echo sequence is

$$\mathbf{H}_{\text{av}} = \pi J2\mathbf{I}_z\mathbf{S}_z, \tag{9.39}$$

where the average Hamiltonian is the Hamiltonian which would give the same total evolution if it was applied to the system for the same total time. This idea of using spin echoes to sculpt the Hamiltonian into some desired average form can be very powerful.

9.7 Evolution of multiple-quantum coherence

The Liouville–von Neumann equation allows us to understand one of the properties of product operators stated without proof in Part A, namely that the zero- and double-quantum coherences of spins I and S do *not* evolve under the J_{IS} coupling (Section 4.3). It is clear from Eqn 8.10 that if the initial density operator and the Hamiltonian commute, then there can be no spin evolution. It is easily verified that all four of the operators involved in these coherences ($2\mathbf{I}_x\mathbf{S}_x$, $2\mathbf{I}_x\mathbf{S}_y$, $2\mathbf{I}_y\mathbf{S}_x$, $2\mathbf{I}_y\mathbf{S}_y$, Eqn 4.5) commute with the weak coupling Hamiltonian $2\pi J\,\mathbf{I}_z\mathbf{S}_z$. This may be done by multiplying the matrix representations of the operators (Appendix I), or by using Eqn 7.29 and its cyclic permutations. For example,

$$\begin{aligned} \left[\mathbf{I}_x\mathbf{S}_y, \mathbf{I}_z\mathbf{S}_z\right] &= \mathbf{I}_x\mathbf{S}_y\mathbf{I}_z\mathbf{S}_z - \mathbf{I}_z\mathbf{S}_z\mathbf{I}_x\mathbf{S}_y \\ &= \mathbf{I}_x\mathbf{I}_z\mathbf{S}_y\mathbf{S}_z - \mathbf{I}_z\mathbf{I}_x\mathbf{S}_z\mathbf{S}_y \\ &= \left(-\tfrac{1}{2}i\mathbf{I}_y\right)\left(\tfrac{1}{2}i\mathbf{S}_x\right) - \left(\tfrac{1}{2}i\mathbf{I}_y\right)\left(-\tfrac{1}{2}i\mathbf{S}_x\right), \\ &= \tfrac{1}{4}\mathbf{I}_y\mathbf{S}_x - \tfrac{1}{4}\mathbf{I}_y\mathbf{S}_x \\ &= 0 \end{aligned} \tag{9.40}$$

where we have used the fact that all I operators commute with all S operators. Appendix L gives a full list of the commutation relations for a two-spin system.

9.8 Summary

- Density operators can be used to describe two-spin systems.
- Density matrices for two-spin systems can be easily worked out using direct products of one-spin matrices.
- Off-diagonal elements in these matrices can be related to different coherence orders.

- The J-coupling interaction has to be considered in combination with the main Zeeman interaction between the spins and the magnetic field.
- Weak coupling occurs when J-coupling constants are small compared with the difference between the frequencies of the coupled spins, corresponding to first-order perturbation theory.
- The weak coupling Hamiltonian commutes with the Zeeman Hamiltonian, which greatly simplifies calculations.
- Total propagators can be calculated for pulse sequences such as spin echoes, and can be interpreted in terms of average Hamiltonians.
- A homonuclear spin echo effectively deletes the Zeeman terms from the average Hamiltonian, leaving only the J-coupling interaction.
- Zero- and double-quantum coherences do not evolve under the corresponding J-coupling interaction.

9.9 Exercises

9.1 Use direct products to calculate the two-spin operator matrix $\boldsymbol{S_x}$ and show that it acts on the four basis states ($\boldsymbol{\psi}_{\alpha\alpha}$ and so on) as expected.

Worked solutions to the exercises are available on the Online Resource Centre at www.oxfordtextbooks.co.uk/orc/hore2e/

9.2 Confirm that the relationships in Eqns 7.28–7.30 hold for spin S in a two-spin system.

9.3 Use a computer package such as *Mathematica* to calculate the matrix representations of the three-spin operators in Section 4.4.

9.4 Repeat the calculations in Eqn 9.16–9.20 for a 90°_y pulse.

9.5 Use a computer package such as *Mathematica* to repeat the double spin echo calculations in Section 9.6. What happens if the 180° pulses are applied with phases of $+x$ and $-x$? How about $+y$ and $-y$?

9.6 Use the methods in Section 9.6 to calculate the effects of a spin echo in a heteronuclear spin system. Devise pulse sequences to produce the three different average Hamiltonians $\Omega_I \boldsymbol{I_z}$, $\Omega_S \boldsymbol{S_z}$, and $\pi J 2\boldsymbol{I_z}\boldsymbol{S_z}$.

9.7 Repeat the calculations in Eqn 9.40 for the other three product operators involved in two-spin multiple quantum coherences.

10 Strong coupling and equivalence

10.1 Introduction

Spin systems comprising weakly coupled spins are straightforward to describe quantum mechanically, as we saw in Chapter 9, because the evolution induced by the various interactions can be calculated sequentially, and in any order. This is no longer true for strong coupling because the coupling term $2\pi J\boldsymbol{I}\cdot\boldsymbol{S}$ does not commute with the other parts of the spin Hamiltonian. For this reason, product operators are useless and there is no option but to use the matrix methods in Chapters 7–9.

We start by calculating the free induction decay for a pair of strongly coupled spin-$\frac{1}{2}$ nuclei excited by a 90° pulse, and then extend the treatment to a spin echo. These calculations illustrate how one might approach more complicated problems. We also use these results to show how the weak coupling approximation emerges in the limit that J is small compared with the frequency difference between the spins. After this we explore the opposite limit, where the coupled spins have precisely the same offset frequency, and describe the phenomenon of *equivalence*. Finally we show how spin locking can be used to make spins *effectively equivalent*, a phenomenon underlying the TOCSY experiment.

10.2 Free induction decay

The full Hamiltonian $\boldsymbol{H_{IS}}$ for a pair of J-coupled spins I and S has been given in Eqns 9.11 and 9.12. Its eigenvalues and eigenvectors are easily obtained (Appendix D):

$$\begin{aligned}
&\lambda_1/2\pi = +\nu+\tfrac{1}{4}J \qquad &&\lambda_2/2\pi = +\tfrac{1}{2}\varepsilon-\tfrac{1}{4}J\\
&\lambda_3/2\pi = -\tfrac{1}{2}\varepsilon-\tfrac{1}{4}J \qquad &&\lambda_4/2\pi = -\nu+\tfrac{1}{4}J
\end{aligned} \tag{10.1}$$

$$\begin{aligned}
|\psi_1\rangle &= |\alpha_I\alpha_S\rangle\\
|\psi_2\rangle &= \cos\theta|\alpha_I\beta_S\rangle+\sin\theta|\beta_I\alpha_S\rangle\\
|\psi_3\rangle &= -\sin\theta|\alpha_I\beta_S\rangle+\cos\theta|\beta_I\alpha_S\rangle\\
|\psi_4\rangle &= |\beta_I\beta_S\rangle
\end{aligned} \tag{10.2}$$

These expressions may be verified by checking that $\hat{H}|\psi_2\rangle = \lambda_2|\psi_2\rangle$, for example.

where

$$2\pi\nu = \tfrac{1}{2}(\Omega_I + \Omega_S), \qquad \varepsilon = \sqrt{J^2 + \delta^2}, \qquad 2\pi\delta = \Omega_I - \Omega_S, \tag{10.3}$$

We use ordinary frequencies here, rather than angular frequencies, so that the same units are used for offset frequencies and coupling constants. The underlying offset frequencies, Ω_I and Ω_S, are still angular frequencies.

and the *mixing angle* θ is given by

$$\tan 2\theta = \frac{J}{\delta}, \quad \text{or} \quad J = \varepsilon \sin 2\theta \quad \text{and} \quad \delta = \varepsilon \cos 2\theta. \tag{10.4}$$

When $|J| \ll |\delta|$, θ is almost zero and the eigenstates reduce to those for weak coupling, i.e. $|\alpha_I\alpha_S\rangle$, $|\alpha_I\beta_S\rangle$, $|\beta_I\alpha_S\rangle$, and $|\beta_I\beta_S\rangle$, with eigenvalues as in Eqn 9.15.

We calculate first the y-component of the free induction decay following a 90°_x pulse:

$$\langle \hat{F}_y \rangle = \mathrm{Tr}\left[\mathbf{F_y}\boldsymbol{\rho}(t)\right] \tag{10.5}$$

We use the total y-magnetization here because the eigenstates $|\psi_2\rangle$ and $|\psi_3\rangle$ (Eqn 10.2) are no longer simple states of the individual spins I and S.

where $\mathbf{F_y}$ is the matrix representation of the operator for the total y-magnetization,

$$\mathbf{F_y} = \mathbf{I_y} + \mathbf{S_y} \tag{10.6}$$

and the density operator is given by

$$\boldsymbol{\rho}(t) = e^{-i\mathbf{H_{IS}}t} e^{-i\frac{1}{2}\pi\mathbf{F_x}} \boldsymbol{\rho}(0) e^{+i\frac{1}{2}\pi\mathbf{F_x}} e^{+i\mathbf{H_{IS}}t} \tag{10.7}$$

cf. Eqn 8.20.

with initial state

$$\boldsymbol{\rho}(0) = \mathbf{I_z} + \mathbf{S_z} = \mathbf{F_z}. \tag{10.8}$$

The x component of the free induction decay can then be calculated in the same way, and the two components combined to give a complex signal. As described in Section 2.2, the combination

Alternatively we could calculate the traditional combination $\langle \hat{F}_x \rangle + i\langle \hat{F}_y \rangle$ and then rephase the spectrum.

$$-\langle \hat{F}_y \rangle + i\langle \hat{F}_x \rangle \tag{10.9}$$

will give a spectrum containing the desired absorption phase lineshapes.

The effect of the pulse is easy to calculate, giving the state $-\mathbf{F_y}$. The exponential operators for the period of free evolution under $\mathbf{H_{IS}}$ after the pulse are more complex to calculate, although the problem can be simplified by noticing that the behaviour of two of the eigenstates, $|\psi_1\rangle$ and $|\psi_4\rangle$, is easy to calculate, leaving a central 2×2 block.The result is

As usual we assume throughout this and the following section that frequency offsets and couplings can be neglected during the pulse. For calculation of the subsequent free evolution, computer assistance is extremely helpful.

$$\mathbf{U} = e^{-i\mathbf{H_{IS}}t} = \begin{pmatrix} e^{-i\pi(J+4\nu)t/2} & 0 & 0 & 0 \\ 0 & U_{22} & U_{23} & 0 \\ 0 & U_{32} & U_{33} & 0 \\ 0 & 0 & 0 & e^{-i\pi(J-4\nu)t/2} \end{pmatrix} \tag{10.10}$$

with

$$\begin{aligned} U_{22} &= e^{i\pi Jt/2}\left[\cos(\pi\varepsilon t) - i\cos(2\theta)\sin(\pi\varepsilon t)\right] \\ U_{23} &= U_{32} = -i\sin(2\theta)e^{i\pi Jt/2}\sin(\pi\varepsilon t) \\ U_{33} &= e^{i\pi Jt/2}\left[\cos(\pi\varepsilon t) + i\cos(2\theta)\sin(\pi\varepsilon t)\right] \end{aligned} \tag{10.11}$$

and the matrix $\mathbf{U}^\dagger$ is easily obtained.

Putting everything together gives:

$$\langle \hat{F}_y \rangle = \mathrm{Tr}\left[\boldsymbol{F_y U}(-\boldsymbol{F_y})\boldsymbol{U}^\dagger \right] \quad \langle \hat{F}_x \rangle = \mathrm{Tr}\left[\boldsymbol{F_x U}(-\boldsymbol{F_y})\boldsymbol{U}^\dagger \right]. \tag{10.12}$$

All we have to do now is to multiply the matrices together, take the trace, and combine the two expectation values together. The result is

$$-\langle \hat{F}_y \rangle + \mathrm{i}\langle \hat{F}_x \rangle = e^{\mathrm{i}2\pi v t} \times 2\left[\cos(\pi J t)\cos(\pi \varepsilon t) + \sin(\pi J t)\sin(\pi \varepsilon t)\sin(2\theta)\right]. \tag{10.13}$$

The initial exponential term indicates a group of lines centred on the frequency $2\pi v$, the average offset frequency of the two spins, while the term within square brackets can be expanded to give the pattern of splittings and intensities. Replacing the cosine and sine terms by complex exponentials gives

$$-\langle \hat{F}_y \rangle + \mathrm{i}\langle \hat{F}_x \rangle = \tfrac{1}{2}\Big[\mathrm{e}^{\mathrm{i}\omega_{13}t}(1-\sin 2\theta) + \mathrm{e}^{\mathrm{i}\omega_{24}t}(1+\sin 2\theta) + \mathrm{e}^{\mathrm{i}\omega_{12}t}(1+\sin 2\theta) + \mathrm{e}^{\mathrm{i}\omega_{34}t}(1-\sin 2\theta) \Big] \tag{10.14}$$

with

As in Eqn 9.22, $\omega_{12} = \lambda_1 - \lambda_2$ is the difference between the eigenvalues for the states $|\psi_1\rangle$ and $|\psi_2\rangle$, and so on.

$$\omega_{13} = 2\pi v + \pi\varepsilon + \pi J \qquad \omega_{24} = 2\pi v + \pi\varepsilon - \pi J$$
$$\omega_{12} = 2\pi v - \pi\varepsilon + \pi J \qquad \omega_{34} = 2\pi v - \pi\varepsilon - \pi J. \tag{10.15}$$

In other words, the free induction decay contains four lines: one pair of lines is centred at a frequency $\pi\varepsilon$ above $2\pi v$ and the other pair centred at a frequency $\pi\varepsilon$ below $2\pi v$, with each pair split by $2\pi J$ so that the lines occur at $\pm\pi J$ from the centre frequencies. If J and δ are positive then the *inner* two lines, at ω_{24} and ω_{12}, have high intensity, $1+\sin 2\theta$, while the *outer* two lines, at ω_{13} and ω_{34}, have low intensity, $1-\sin 2\theta$. This NMR spectrum from a strongly coupled pair of spins is shown in Fig. 10.1.

If either J or δ is negative then the lines swap positions *and* amplitudes, so that the spectrum looks exactly the same; if both are negative then the spectrum again looks exactly the same.

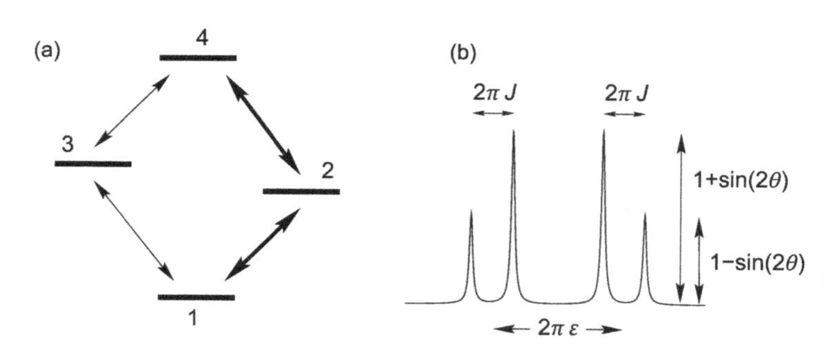

Fig. 10.1 Energy levels (a) and schematic spectrum (b) for a strongly coupled pair of spin-$\frac{1}{2}$ nuclei. The thick and thin arrows indicate the more allowed and less allowed NMR transitions, respectively.

Note that even for strongly coupled spins the two doublets continue to be split by $2\pi J$, but the doublets are no longer centred at Ω_I and Ω_S. The physics behind the line shifts and intensity changes predicted for strong coupling lies in the mixing of the $|\alpha_I\beta_S\rangle$ and $|\beta_I\alpha_S\rangle$ states (Eqn 10.2) and the consequent modification of the NMR transition probabilities. The outer lines become 'less allowed', or 'more forbidden', while the inner lines become 'more allowed'. In the weak coupling limit,

$$|2\pi J| \ll |\Omega_I - \Omega_S| \quad \Rightarrow \quad |J| \ll |\delta| \quad \Rightarrow \quad \varepsilon \approx \delta \quad \Rightarrow \quad \theta \approx 0, \tag{10.16}$$

and so the four amplitudes become identical, while the doublets, with splitting $2\pi J$, are now centred at the chemical shift positions, Ω_I and Ω_S.

This reproduces the results of Eqn 9.23.

Doing all this algebra swiftly becomes tiresome, even with computer assistance, and a more intuitive feel for the behaviour of strong couplings can be obtained by using computers to calculate the final spectrum directly and numerically. Figure 10.2 shows an example calculation in *Mathematica*. It is easy to vary the parameters to explore the range of spectra which can be obtained.

Computer programs such as *Mathematica* are discussed in Appendix D.

10.3 Spin echoes

We now consider the echo sequence $90^\circ_x - \tau - 180^\circ_x - \tau$, applied to an initial thermal equilibrium state, which we previously considered within the weak coupling approximation in Section 9.6. The echo signal is given by the trace of $\boldsymbol{F_y}$ multiplied by $\boldsymbol{\rho}(2\tau)$, which is obtained by allowing the density operator to evolve sequentially during the four parts of the experiment.

In Section 9.6 we in fact considered the more general problem of applying the echo to *any* initial state.

Following the same procedure as before

$$\langle \hat{F}_y \rangle = \mathrm{Tr}\left[\boldsymbol{F_y U} \mathrm{e}^{-\mathrm{i}\pi \boldsymbol{F_x}} \boldsymbol{U} \mathrm{e}^{-\mathrm{i}\frac{1}{2}\pi \boldsymbol{F_x}} \boldsymbol{F_z} \mathrm{e}^{+\mathrm{i}\frac{1}{2}\pi \boldsymbol{F_x}} \boldsymbol{U}^\dagger \mathrm{e}^{+\mathrm{i}\pi \boldsymbol{F_x}} \boldsymbol{U}^\dagger \right] \tag{10.17}$$

where the state after the initial 90°_x pulse is easily calculated to be $-\boldsymbol{F_y}$, $\boldsymbol{U}$ is given by Eqns 10.10 and 10.11, and the propagator for the 180° pulse was previously calculated in Eqn 9.35. After a lot of matrix multiplication, one comes to the result:

$$\begin{aligned} \langle \hat{F}_y \rangle = {} & 2\cos^2 2\theta \cos 2\pi J\tau \\ & - \sin 2\theta (1 - \sin 2\theta) \cos(2\pi[J + \varepsilon]\tau) \\ & + \sin 2\theta (1 + \sin 2\theta) \cos(2\pi[J - \varepsilon]\tau). \end{aligned} \tag{10.18}$$

Clearly, the echo modulation for a strongly coupled pair of spins is more complicated than for weak coupling: we have three frequencies with different amplitudes that depend on the ratio J/δ, i.e. the strength of the coupling. The weak coupling limit is easily obtained, as the mixing angle $\theta \approx 0$, and the second and third terms disappear. It is easy to show that

$$\langle \hat{F}_x \rangle = 0 \tag{10.19}$$

and so there is no echo along the x-axis, however weak or strong the coupling.

2-spin system with strong coupling

- **Set up**

```
In[1]:= e = (1 0
             0 1);   σx = (0    1/2
                           1/2  0  );   σy = (0    -i/2
                                              i/2   0  );
        σz = (1/2   0
              0    -1/2); (* single spin operators *)

        M1_ ⊗ M2_ := KroneckerProduct[M1, M2]; (* direct product *)

        Ix = σx ⊗ e; Sx = e ⊗ σx; Iy = σy ⊗ e; Sy = e ⊗ σy;
        Iz = σz ⊗ e; Sz = e ⊗ σz; (* spin operators *)
```

- **Numerical calculation**

```
In[4]:= H = ΩI Iz + ΩS Sz + 2 π J (Ix.Sx + Iy.Sy + Iz.Sz) // Simplify; (* Hamiltonian during fid *)
In[5]:= ρeq = Iz + Sz; (* equilibrium density matrix *)

        ρ = MatrixExp[-i H t].MatrixExp[-i π (Ix + Sx) / 2].
           ρeq.MatrixExp[i π (Ix + Sx) / 2].MatrixExp[i H t];
            (* density matrix during fid, after a 90 degree x-pulse *)

        fid = - (Tr[ρ.(Iy + Sy)] - i Tr[ρ.(Ix + Sx) ]) Exp[-t / T2] // Simplify;
            (* free induction decay *)
            (* i.e. expectation value of  transverse magnetization with exponential spin-
          spin relaxation and quadrature detection *)

In[8]:= f = fid /. {J → 0.5, ΩI → -5.0, ΩS → 5.0, T2 → 10.0};
        (* Values of parameters, arbitrary units *)
        s = FourierTransform[f × UnitStep[t], t, ω]; (* spectrum *)
        pf = Plot[Re[f], {t, 0, 50}, PlotRange → All, Ticks → None];
        ps = Plot[Re[s], {ω, -15, 15}, PlotRange → All, Axes → None];
        GraphicsRow[{pf, ps}]
```

Out[12]=

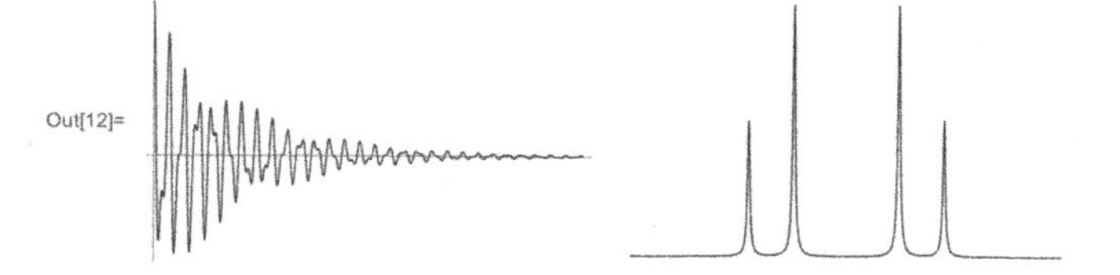

Fig. 10.2 Strong coupling spectrum calculated using *Mathematica*. The effects of relaxation have been simulated by imposing an exponential decay on the calculated free induction decay. It is easy to vary the different parameters and explore the range of spectra which can be produced.

10.4 Equivalent spins

Many molecules contain groups of two or more equivalent nuclei, such as the three protons in a freely rotating methyl group. Even though such nuclei are coupled to one another, no splitting is observed in the NMR spectrum. We can now see why.

By equivalent here we mean *magnetically equivalent*. A set of nuclei (a, b, c,...) with identical chemical shifts is magnetically equivalent either if there are no other spins in the molecule or if, for every other nucleus (e.g. q) in the molecule, the spin-spin coupling constants satisfy the relation $J_{aq}=J_{bq}=J_{cq}=...$ A set of nuclei with identical chemical shifts but different couplings is *chemically equivalent*. See Chapter 3 of Hore (2015) for some examples.

Equivalent spins correspond to the extreme limit of strong coupling, where $\delta=0$ and the mixing angle takes on its maximum value, $\theta=45°$, so

$$\begin{aligned}\sin(2\theta)&=1,\\ \cos(2\theta)&=0.\end{aligned} \tag{10.20}$$

Consider first the case of the free induction decay, calculated in Section 10.2. The outer lines in Fig. 10.1 now have no intensity (they are now completely forbidden), and since $\varepsilon=J$ the two inner lines coincide. Thus *no sign of the coupling appears* in the NMR spectrum. Similarly, in the case of a spin echo (Eqn 10.18) only the third term survives, and this term is now constant, so there is no modulation of the echo by the coupling.

These results can also be understood by considering the eigenstates and eigenvalues of the strong coupling Hamiltonian, after substituting $\theta=45°$ into Eqns 10.1 and 10.2. Three of the eigenstates ($|\psi_1\rangle, |\psi_2\rangle$, and $|\psi_4\rangle$) form a *triplet* of states, separated from each other by gaps of $2\pi\nu$, while the fourth one, $|\psi_3\rangle$, is a *singlet* state. Transitions between the three triplet states are allowed, giving rise to the NMR signal at frequency ν, but no transitions can occur between them and the isolated singlet state.

The three triplet states are symmetric under permutations of the nuclear labels, while the singlet state is antisymmetric; see Atkins and Friedman (2010). Remember that, for equivalent spins, $\varepsilon=J$.

Instead of proceeding via the general case of strong coupling, these results can instead be calculated directly. The brute-force approach is to write down the Hamiltonian (Eqn 9.12) with $\Omega_I=\Omega_S=\Omega$:

$$\boldsymbol{H_{IS}}=\Omega(\boldsymbol{I_z}+\boldsymbol{S_z})+2\pi J\boldsymbol{I}\cdot\boldsymbol{S}=\begin{pmatrix}\Omega+\frac{1}{2}\pi J & 0 & 0 & 0\\ 0 & -\frac{1}{2}\pi J & \pi J & 0\\ 0 & \pi J & -\frac{1}{2}\pi J & 0\\ 0 & 0 & 0 & -\Omega+\frac{1}{2}\pi J\end{pmatrix} \tag{10.21}$$

and calculate its exponential

$$e^{\pm i\boldsymbol{H_{IS}}t}=\begin{pmatrix}e^{\pm i(\Omega+\pi J/2)t} & 0 & 0 & 0\\ 0 & \frac{1}{2}e^{\pm i\pi Jt/2}+\frac{1}{2}e^{\mp 3i\pi Jt/2} & \frac{1}{2}e^{\pm i\pi Jt/2}-\frac{1}{2}e^{\mp 3i\pi Jt/2} & 0\\ 0 & \frac{1}{2}e^{\pm i\pi Jt/2}-\frac{1}{2}e^{\mp 3i\pi Jt/2} & \frac{1}{2}e^{\pm i\pi Jt/2}+\frac{1}{2}e^{\mp 3i\pi Jt/2} & 0\\ 0 & 0 & 0 & e^{\pm i(-\Omega+\pi J/2)t}\end{pmatrix} \tag{10.22}$$

using the method described in Appendix D. Matrix multiplication can then be used to find the density matrix during the free induction decay reproducing the previous results: the spectrum will comprise a single line at the chemical shift frequency Ω and *no* splitting due to J-coupling.

This calculation is not fundamentally different from the full strong coupling calculation above, but making the simplifying assumption at the start of the calculation makes the maths much simpler.

There is, however, a more elegant and general method. Consider a spin system described by a Hamiltonian $\boldsymbol{H}$ comprising two parts with the following properties:

$$\boldsymbol{H}=\boldsymbol{H}_1+\boldsymbol{H}_2 \quad \text{and} \quad [\boldsymbol{H}_1,\boldsymbol{H}_2]=[\boldsymbol{F}_x,\boldsymbol{H}_2]=[\boldsymbol{F}_y,\boldsymbol{H}_2]=0 \tag{10.23}$$

So $\boldsymbol{F}_x=\boldsymbol{I}_x+\boldsymbol{S}_x$ for a two-spin system, as before.

where $\boldsymbol{F}_x$ and $\boldsymbol{F}_y$ are the total x and y operators, summed over all spins. The observable x-magnetization is given by

These expressions require some of the properties of exponential matrices outlined in Appendix K. In going from line 1 to line 2 we have used the fact that the trace is invariant to cyclic permutation of the operators within it (Appendix B).

$$\begin{aligned}\mathrm{Tr}\left[\boldsymbol{F}_x e^{-i\boldsymbol{H}t}\boldsymbol{\rho}(0)e^{+i\boldsymbol{H}t}\right] &= \mathrm{Tr}\left[\boldsymbol{F}_x e^{-i\boldsymbol{H}_2 t}e^{-i\boldsymbol{H}_1 t}\boldsymbol{\rho}(0)e^{+i\boldsymbol{H}_1 t}e^{+i\boldsymbol{H}_2 t}\right]\\ &= \mathrm{Tr}\left[e^{+i\boldsymbol{H}_2 t}\boldsymbol{F}_x e^{-i\boldsymbol{H}_2 t}e^{-i\boldsymbol{H}_1 t}\boldsymbol{\rho}(0)e^{+i\boldsymbol{H}_1 t}\right]\\ &= \mathrm{Tr}\left[\boldsymbol{F}_x e^{+i\boldsymbol{H}_2 t}e^{-i\boldsymbol{H}_2 t}e^{-i\boldsymbol{H}_1 t}\boldsymbol{\rho}(0)e^{+i\boldsymbol{H}_1 t}\right]\\ &= \mathrm{Tr}\left[\boldsymbol{F}_x e^{-i\boldsymbol{H}_1 t}\boldsymbol{\rho}(0)e^{+i\boldsymbol{H}_1 t}\right].\end{aligned} \tag{10.24}$$

The equivalent result holds for the y component of magnetization. Thus $\boldsymbol{H}_2$ has no observable effect, whatever the initial state $\boldsymbol{\rho}(0)$.

To see how all this relates to equivalence, consider a system of four spins, I_1, I_2, S, and R in which I_1 and I_2 are equivalent. $\boldsymbol{H}_2$ is the coupling term between the I spins:

$$\boldsymbol{H}_2 = 2\pi J_{12}\boldsymbol{I}_1\cdot\boldsymbol{I}_2 \tag{10.25}$$

and it commutes with all the other terms in the Hamiltonian:

The IS subscript on J_{IS} is briefly reinstated here to avoid confusion.

$$\begin{aligned}\boldsymbol{H}_1 = {}&\Omega_\mathrm{I}(\boldsymbol{I}_{1z}+\boldsymbol{I}_{2z})+\Omega_\mathrm{S}\boldsymbol{S}_z+\Omega_\mathrm{R}\boldsymbol{R}_z+2\pi J_{\mathrm{IS}}(\boldsymbol{I}_1+\boldsymbol{I}_2)\cdot\boldsymbol{S}\\ &+2\pi J_{\mathrm{IR}}(\boldsymbol{I}_1+\boldsymbol{I}_2)\cdot\boldsymbol{R}+2\pi J_{\mathrm{SR}}\boldsymbol{S}\cdot\boldsymbol{R}.\end{aligned} \tag{10.26}$$

$\boldsymbol{H}_2$ also commutes with the total x and y operators, e.g. $\boldsymbol{F}_x=\boldsymbol{I}_{1x}+\boldsymbol{I}_{2x}+\boldsymbol{S}_x+\boldsymbol{R}_x$, and therefore has no effect on the observed signal. The same result is *not* true for inequivalent spins because the coupling term does not commute with the Zeeman interactions of the I spins:

$$\left[2\pi J_{12}\boldsymbol{I}_1\cdot\boldsymbol{I}_2,(\Omega_1\boldsymbol{I}_{1z}+\Omega_2\boldsymbol{I}_{2z})\right]\neq\boldsymbol{0}. \tag{10.27}$$

This approach is easily generalized to more than two equivalent spins and more than one group of equivalent spins.

10.5 TOCSY

We are at last in a position to justify the expressions given in Eqn 5.13 for the evolution during a period of spin locking, as in the TOCSY experiment (Section 5.5). The Hamiltonian in the rotating frame for an IS spin system, with a spin-locking field along the x-axis, is

$$\hat{H}=\omega_{1\mathrm{I}}\hat{I}_x+\omega_{1\mathrm{S}}\hat{S}_x+\Omega_\mathrm{I}\hat{I}_z+\Omega_\mathrm{S}\hat{S}_z+2\pi J\hat{I}\cdot\hat{S} \tag{10.28}$$

where $\omega_{1\mathrm{I}}$ and $\omega_{1\mathrm{S}}$ are the radiofrequency field strengths experienced by the two spins. For strong spin-locking fields ($\omega_{1\mathrm{I}}\gg\Omega_\mathrm{I}$, $\omega_{1\mathrm{S}}\gg\Omega_\mathrm{S}$), both spins experience an effective field along the x-axis, so that the Zeeman terms in Eqn 9.36 can be dropped. In a homonuclear spin system, such as a pair of ^{1}H nuclei, ω_{1I} and ω_{1S} are almost identical; they differ by a few parts per million, but since $\omega_1/2\pi$ is typically about 10 kHz this difference will be only a tiny fraction of a Hz. Thus,

$$\hat{H} \approx \omega_1\left(\hat{I}_x + \hat{S}_x\right) + 2\pi J \hat{I} \cdot \hat{S} \tag{10.29}$$

which can be seen to have exactly the same form as the Hamiltonian for two equivalent spins (Eqn 10.21) except that Ω is replaced by ω_1, and z by x. In a heteronuclear spin system different radiofrequency fields are used to irradiate the two spins, so that ω_{1I} and ω_{1S} can be separately controlled. In particular, the two frequencies can be chosen to be the same (the Hartmann–Hahn condition), once again making the spins effectively equivalent.

Since the two parts of Eqn 10.29 commute, we can treat them sequentially and in any order. The first term has no effect on either $\boldsymbol{I_x}$ or $\boldsymbol{S_x}$ because they both commute with $\boldsymbol{F_x}=\boldsymbol{I_x}+\boldsymbol{S_x}$. It is the J-coupling that produces the desired effect. Using Eqn 8.51 with $\boldsymbol{A}=(\boldsymbol{I_x}-\boldsymbol{S_x})$, $\boldsymbol{B}=\boldsymbol{I}\cdot\boldsymbol{S}$, $\boldsymbol{C}=(-2\boldsymbol{I_y}\boldsymbol{S_z}+2\boldsymbol{I_z}\boldsymbol{S_y})$, and $b=2\pi J$, one finds (cf. Eqn 5.13)

$$\boldsymbol{I_x} - \boldsymbol{S_x} \rightarrow (\boldsymbol{I_x} - \boldsymbol{S_x})\cos(2\pi Jt) + \left(2\boldsymbol{I_y}\boldsymbol{S_z} - 2\boldsymbol{I_z}\boldsymbol{S_y}\right)\sin(2\pi Jt). \tag{10.30}$$

$\boldsymbol{I}\cdot\boldsymbol{S}$ does, however, commute with $\boldsymbol{I_x}+\boldsymbol{S_x}$, so that the sum of the x operators does not evolve. Similar effects follow for $\boldsymbol{I_y}\pm\boldsymbol{S_y}$ and $\boldsymbol{I_z}\pm\boldsymbol{S_z}$.

The $\hat{I}\cdot\hat{S}$ term is completely symmetric in its actions along the x, y, and z axes.

Any in-phase y- and z-magnetizations will also evolve under the first term in Eqn 10.29, e.g.

$$\boldsymbol{I_y} \rightarrow \boldsymbol{I_y}\cos\omega_1 t + \boldsymbol{I_z}\sin\omega_1 t. \tag{10.31}$$

The B_1 field strength, and thus ω_1, will vary across the sample, and since the spin-locking field is applied for a long time this will act to dephase $\boldsymbol{I_y}$ and $\boldsymbol{I_z}$ terms. $\boldsymbol{I_x}$ terms will not be dephased, because they commute with the $\boldsymbol{I_x}$ part of the Hamiltonian and so will not evolve under it.

The evolution of antiphase terms is more complex. The difference term in Eqn 10.30 commutes with $\boldsymbol{F_x}$,

$$\left[2\boldsymbol{I_y}\boldsymbol{S_z} - 2\boldsymbol{I_z}\boldsymbol{S_y}, \boldsymbol{I_x} + \boldsymbol{S_x}\right] = 0, \tag{10.32}$$

and so is not dephased, but the sum term does not commute with $\boldsymbol{F_x}$ and so will be dephased. The simple antiphase y-magnetization term $2\boldsymbol{I_y}\boldsymbol{S_z}$ can always be rewritten as a linear combination of sum and difference terms

$$2\boldsymbol{I_y}\boldsymbol{S_z} = \tfrac{1}{2}\left(2\boldsymbol{I_y}\boldsymbol{S_z} - 2\boldsymbol{I_z}\boldsymbol{S_y}\right) + \tfrac{1}{2}\left(2\boldsymbol{I_y}\boldsymbol{S_z} + 2\boldsymbol{I_z}\boldsymbol{S_y}\right), \tag{10.33}$$

and so will be partially dephased. By contrast,the antiphase x-magnetization, i.e. terms of the form $2\boldsymbol{I_x}\boldsymbol{S_z}$, is completely dephased.

For genuinely equivalent spins the evolution described by Eqn 10.30 can have no discernible effect, as the *observable* magnetization does not evolve under the J-coupling (Section 10.4). Spins which are rendered temporarily equivalent, however, can show much more complex behaviour, as they can enter the spin-lock period in a variety of initial states, and can be observed as inequivalent spins at the end of the spin-locking period.

10.6 Summary

- The full strong coupling Hamiltonian does not commute with the Zeeman Hamiltonian, making calculations complicated.
- Two-spin systems are simple enough for calculations by hand, but computer assistance is very useful.
- The strong coupling mixes two of the spin states.
- This mixing results in a spectrum where the inner two lines grow in intensity and the outer lines shrink.
- The splitting within each doublet is still given by the *J*-coupling constant, but the line positions are altered.
- Spin echoes still work, but the modulation by the *J*-coupling is more complex than for weak coupling.
- Weak coupling behaviour emerges in the limit of small coupling.
- Equivalent spin behaviour occurs when the coupled spins have exactly the same frequency.
- In this limit the calculations are also relatively simple, and the *J*-coupling has no visible effect.
- The mixing of spin states is now complete to give a triplet of spin states and an isolated singlet state.
- The TOCSY pulse sequence uses spin locking to render two spins temporarily equivalent, allowing the effects of the coupling to be observed after the spin lock is switched off.

10.7 Exercises

Worked solutions to the exercises are available on the Online Resource Centre at www.oxfordtextbooks.co.uk/orc/hore2e/

10.1. Explain why the eigenvectors of $\boldsymbol{H}_{IS}$ must have the general form shown in Eqn 10.2, for some value of θ. (Two eigenvectors can be spotted immediately, and then use the fact that the remaining eigenvectors of $\boldsymbol{H}_{IS}$ will be orthogonal unit vectors.)

10.2. Verify the eigenvalues listed in Eqn 10.1.

10.3. Use λ_2 and $|\psi_2\rangle$ to find an expression for $\tan\theta$ and then use double angle formulae to verify the expression for $\tan 2\theta$.

10.4. Use a computer package such as *Mathematica* to verify Eqns 10.14 and 10.15 directly.

10.5. Use your program to explore the spectrum for a range of values of the ratio J/δ and convince yourself that weak coupling and equivalent spin behaviour emerges in the appropriate limits.

10.6. Expand your program to verify Eqns 10.18 and 10.19.

10.7. Use matrix representations to prove Eqn 10.32.

10.8. Use operator commutators to prove Eqn 10.32, and hence evaluate the commutators between the sum term $2\boldsymbol{I}_y\boldsymbol{S}_z+2\boldsymbol{I}_z\boldsymbol{S}_y$ and $\boldsymbol{F}_x$.

Appendices

Appendix A. NOE, cross relaxation, and the Solomon equations

The important part of a NOESY experiment—the bit when things really happen—is the mixing period τ_m. During this interval, cross relaxation and/or slow chemical exchange transfer some of the intensity of the modulated I_z state into S_z, carrying with it the modulation at the offset frequency Ω_I. We can see how cross relaxation brings about this transfer by considering the energy-level diagram for a two-spin IS system (Fig. A1).

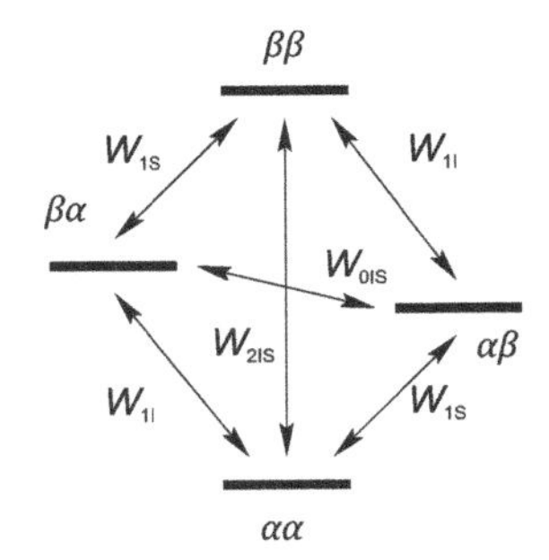

Fig. A1 Energy-level diagram for an IS spin system with the six possible cross relaxation pathways.

Spin-lattice or longitudinal relaxation is caused by fluctuating magnetic fields in a liquid sample which induce transitions between spin states. If these fields arise from the dipolar interaction of the two spins, then all six possible transitions occur. The relaxation rate constants are labelled W_{1I} if the I spin flips and the magnetic quantum number m changes by $\Delta m = \pm 1$, W_{1S} if the S-spin flips and $\Delta m = \pm 1$, W_{2IS} if both I and S flip in the same sense such that $\Delta m = \pm 2$, and W_{0IS} if I and S flip in opposite senses and $\Delta m = 0$.

The effect of these spin flips is to move the populations of the four energy levels towards equilibrium. Ignoring the equilibrium population differences for simplicity for the moment, we have the rate equations:

$$\frac{dn_{\alpha\alpha}}{dt} = -(W_{1I} + W_{1S} + W_{2IS})n_{\alpha\alpha} + W_{1I}n_{\beta\alpha} + W_{1S}n_{\alpha\beta} + W_{2IS}n_{\beta\beta}$$
$$\frac{dn_{\alpha\beta}}{dt} = -(W_{1I} + W_{1S} + W_{0IS})n_{\alpha\beta} + W_{1I}n_{\beta\beta} + W_{1S}n_{\alpha\alpha} + W_{0IS}n_{\beta\alpha} \quad \text{(A1)}$$

$n_{\alpha\alpha}$, $n_{\alpha\beta}$, $n_{\beta\alpha}$, and $n_{\beta\beta}$ are the populations of the four energy levels shown in Fig. A1.

and similarly for $dn_{\beta\alpha}/dt$ and $dn_{\beta\beta}/dt$. After some tedious but straightforward rearrangement, one can obtain the two coupled rate equations

$$\frac{d(n_{\alpha\alpha} + n_{\alpha\beta} - n_{\beta\alpha} - n_{\beta\beta})}{dt} = -\rho_I(n_{\alpha\alpha} + n_{\alpha\beta} - n_{\beta\alpha} - n_{\beta\beta}) - \sigma_{IS}(n_{\alpha\alpha} - n_{\alpha\beta} + n_{\beta\alpha} - n_{\beta\beta})$$
$$\frac{d(n_{\alpha\alpha} - n_{\alpha\beta} + n_{\beta\alpha} - n_{\beta\beta})}{dt} = -\sigma_{IS}(n_{\alpha\alpha} + n_{\alpha\beta} - n_{\beta\alpha} - n_{\beta\beta}) - \rho_S(n_{\alpha\alpha} - n_{\alpha\beta} + n_{\beta\alpha} - n_{\beta\beta}) \quad \text{(A2)}$$

where

$$\rho_I = W_{0IS} + 2W_{1I} + W_{2IS}$$
$$\rho_S = W_{0IS} + 2W_{1S} + W_{2IS}$$
$$\sigma_{IS} = W_{2IS} - W_{0IS}. \quad \text{(A3)}$$

But $(n_{\alpha\alpha} + n_{\alpha\beta} - n_{\beta\alpha} - n_{\beta\beta})$ is just the total number of I spins in the α state minus the total number in the β state; $(n_{\alpha\alpha} + n_{\alpha\beta} - n_{\beta\alpha} - n_{\beta\beta})$ is therefore proportional to the z component of the I-spin magnetization. Similarly, $(n_{\alpha\alpha} - n_{\alpha\beta} + n_{\beta\alpha} - n_{\beta\beta})$ is proportional to the z component of the S-spin magnetization. Defining $a_I(t)$ and $a_S(t)$ as the amplitudes of the I_z and S_z operators at time t, we get

$$\frac{da_I}{dt} = -\rho_I a_I - \sigma_{IS} a_S$$
$$\frac{da_S}{dt} = -\sigma_{IS} a_I - \rho_S a_S. \qquad \text{(A4)}$$

To ensure that relaxation restores the (Boltzmann) equilibrium magnetizations, we can simply replace a_I and a_S on the right hand side by $(a_I - a_I^0)$ and $(a_S - a_S^0)$ where the superscript zero indicates the value at equilibrium:

This would not have been necessary if we had included the equilibrium populations at the outset.

$$\frac{da_I}{dt} = -\rho_I (a_I - a_I^0) - \sigma_{IS} (a_S - a_S^0)$$
$$\frac{da_S}{dt} = -\sigma_{IS} (a_I - a_I^0) - \rho_S (a_S - a_S^0). \qquad \text{(A5)}$$

Solution of these coupled equations gives the time-dependence of the coefficients $a_I(t)$ and $a_S(t)$, and hence the evolution of the z-magnetization during the NOESY mixing period. The expressions are known as the Solomon equations. They show that z-magnetization will be transferred between the two spins *provided* the difference between the two cross relaxation rates, σ_{IS}, is non-zero.

The Solomon equations may also be used to understand the nuclear Overhauser effect (NOE) more generally. For example, the form of the σ_{IS} term (Eqn A3), without which there would be no magnetization transfer, shows clearly why the sign of the NOE depends on whether W_{2IS} is faster or slower than W_{0IS}, as discussed in Hore (2015).

Appendix B. Vectors and matrices

A *matrix* is an array of *scalar* numbers arranged in rows and columns, e.g.

$$\boldsymbol{M} = \begin{pmatrix} p & q \\ r & s \end{pmatrix}. \qquad \text{(B1)}$$

The individual numbers making up a matrix are called *elements*, and are addressed according to their position in the matrix by stating first their *row* number and then their *column* number; these are normally written as subscripts and so, for example, we write $M_{12} = q$. Note that the name of a matrix is printed in ***bold italic*** type, while the individual elements are printed in *italic* type.

Complex numbers can be written as $z = a + ib$, where $i = \sqrt{-1}$, or equivalently as $z = re^{i\theta}$; here a, b, r, and θ are *real*.

Many simple introductions to matrices implicitly assume that the individual matrix elements are *real* numbers. In fact most matrices used in quantum mechanics have *complex* elements but this does not greatly alter the discussion:

the basic rules for manipulating complex matrices are either identical to or very similar to the corresponding rules for real matrices.

If a matrix has only a single row it is called a *row vector*; if it has only a single column it is called a *column vector*, or just a *vector*. For example,

$$\boldsymbol{a} = \begin{pmatrix} a_x \\ a_y \\ a_z \end{pmatrix} \tag{B2}$$

is a vector with three elements, the x, y, and z components of $\boldsymbol{a}$.

Two matrices, $\boldsymbol{M}$ and $\boldsymbol{N}$, may be added or multiplied as follows. The sum, $\boldsymbol{S} = \boldsymbol{M} + \boldsymbol{N}$, has elements given by

$$S_{rc} = M_{rc} + N_{rc}, \tag{B3}$$

while the product, $\boldsymbol{P} = \boldsymbol{MN}$, has

$$P_{rc} = \sum_j M_{rj} N_{jc}. \tag{B4}$$

It is only possible to add two matrices if they have the same shape, i.e. the same number of rows and columns. Similarly, it is only possible to multiply two matrices if the first matrix has as many columns as the second has rows. Matrix addition is *commutative*, i.e. $\boldsymbol{M} + \boldsymbol{N} = \boldsymbol{N} + \boldsymbol{M}$, but matrix multiplication is not, and so $\boldsymbol{MN} \neq \boldsymbol{NM}$ in general. Sometimes, however, $\boldsymbol{MN}$ does equal $\boldsymbol{NM}$, and in such cases the two matrices $\boldsymbol{M}$ and $\boldsymbol{N}$ are said to *commute*. Clearly two matrices can only commute if they are both *square* (i.e. have the same number of rows as columns) and of the same size.

Matrix multiplication is, however, *associative*, so that $\boldsymbol{A}(\boldsymbol{BC}) = (\boldsymbol{AB})\boldsymbol{C}$.

Matrix multiplication is *distributive over addition*, so that $\boldsymbol{A}(\boldsymbol{B} + \boldsymbol{C}) = \boldsymbol{AB} + \boldsymbol{AC}$. Similarly, matrix operations are in general *linear*, which means that an operation can be applied to the sum of two matrices by applying the operation to each matrix separately, and then adding the two results. This allows many operations to be simplified by breaking them down into smaller components.

In addition to these *binary* matrix operations, which combine two matrices to produce a third, there are also *unary* operations, which act on a single matrix. One simple example is the matrix *transpose*, defined by interchanging rows and columns; for square matrices this corresponds to swapping the elements around the *principal diagonal*. For our example matrix,

$$\boldsymbol{M}^{\mathrm{T}} = \begin{pmatrix} p & r \\ q & s \end{pmatrix}. \tag{B5}$$

Note that transposition will convert a column vector into a row vector and vice versa. It can be shown that $(\boldsymbol{AB})^{\mathrm{T}} = \boldsymbol{B}^{\mathrm{T}}\boldsymbol{A}^{\mathrm{T}}$; this is easily verified for a 2×2 matrix by direct multiplication. A matrix which is equal to its own transpose is said to be *symmetric*.

All the properties above are defined equally well for both real and complex matrices, but the matrix transpose is rarely a useful operation for complex matrices: the matrix *adjoint* or *Hermitian conjugate* is normally used instead. This is defined as the transpose of the *conjugate* of the matrix (that is, the matrix with elements equal to the *complex conjugates* of the corresponding elements in the transposed matrix). For our example matrix,

N.B. $(a+ib)^* = a-ib$ and $\left(re^{i\theta}\right)^* = re^{-i\theta}$.

$$\boldsymbol{M}^\dagger = \begin{pmatrix} p^* & r^* \\ q^* & s^* \end{pmatrix}. \tag{B6}$$

A matrix which is equal to its own adjoint, $\boldsymbol{M}^\dagger = \boldsymbol{M}$, is said to be *Hermitian*. Note that for a *real* matrix $\boldsymbol{M}^\dagger = \boldsymbol{M}^\mathsf{T}$, so the adjoint is equivalent to the transpose and a real Hermitian matrix is symmetric. The adjoint also has the useful property that

$$(\boldsymbol{AB})^\dagger = \boldsymbol{B}^\dagger \boldsymbol{A}^\dagger \tag{B7}$$

so the adjoint of the product of two matrices is the reversed product of their adjoints.

A final important matrix operation is the *trace*, which converts a square matrix to a number, equal to the sum of the elements on the principal diagonal of the matrix. A useful property of the trace of a *product* of matrices is that it does not change if the matrices are *cyclically permuted*, e.g. $\boldsymbol{ABC} \to \boldsymbol{CAB} \to \boldsymbol{BCA}$. This may easily be demonstrated:

A_{jk}, B_{km}, and C_{mj} are just numbers, and can be multiplied in any order.

$$\begin{aligned} \mathrm{Tr}[\boldsymbol{ABC}] &= \sum_j (\boldsymbol{ABC})_{jj} = \sum_j \sum_k \sum_m A_{jk} B_{km} C_{mj} \\ &= \sum_j \sum_k \sum_m C_{mj} A_{jk} B_{km} = \sum_m (\boldsymbol{CAB})_{mm} = \mathrm{Tr}[\boldsymbol{CAB}] \\ &= \sum_j \sum_k \sum_m B_{km} C_{mj} A_{jk} = \sum_k (\boldsymbol{BCA})_{kk} = \mathrm{Tr}[\boldsymbol{BCA}]. \end{aligned} \tag{B8}$$

The result holds however many matrices are multiplied within the trace, and however many are cyclically permuted.

Turning now to vectors, it is clear that one can add two vectors by adding corresponding elements, as long as the vectors have the same dimension; this is just a special case of matrix addition. Similarly two vectors can be multiplied to form the *vector dot product* or *scalar product* by multiplying corresponding elements and adding them together. For two vectors, $\boldsymbol{a}$ and $\boldsymbol{b}$,

Much of the discussion here assumes that the vectors are conventional three-dimensional vectors, but the ideas can be easily generalized to arbitrary dimensions.

$$\boldsymbol{a} \cdot \boldsymbol{b} = a_x b_x + a_y b_y + a_z b_z. \tag{B9}$$

The dot product can also be viewed as a special case of matrix multiplication,

$$\boldsymbol{a} \cdot \boldsymbol{b} = \boldsymbol{a}^\mathsf{T} \boldsymbol{b}, \tag{B10}$$

Strictly speaking this equation is not quite correct, as the dot product gives a scalar number while the matrix product gives a matrix containing this number as its sole element, but this subtlety is frequently ignored. The equation can be made rigorously correct by writing $\boldsymbol{a} \cdot \boldsymbol{b} = \mathrm{Tr}(\boldsymbol{a}^\mathsf{T}\boldsymbol{b})$, where Tr is the matrix trace discussed above.

but it is sufficiently important to be considered as an operation in its own right. For example, the dot product can be used to determine the length, a, of a vector $\boldsymbol{a}$:

$$a = \sqrt{a_x^2 + a_y^2 + a_z^2} = \sqrt{\boldsymbol{a} \cdot \boldsymbol{a}}. \tag{B11}$$

As usual we write the vector in ***bold italic*** type, and its length in *italic* type. A vector of length 1 is called a *unit vector*.

Similarly for two vectors, $\boldsymbol{a}$ and $\boldsymbol{b}$,

$$\boldsymbol{a}\cdot\boldsymbol{b} = ab\cos\theta, \tag{B12}$$

where θ is the angle between the two vectors. If $\theta = 90°$ then $\boldsymbol{a}\cdot\boldsymbol{b} = 0$, and the two vectors are said to be *orthogonal*.

A very similar approach can be used for complex vectors, but as one might guess the transpose operation should be replaced by the adjoint, and so the *inner product* of two complex vectors is defined by

$$\boldsymbol{a}\cdot\boldsymbol{b} = \boldsymbol{a}^{\dagger}\boldsymbol{b} \tag{B13}$$

which reduces to the conventional definition (Eqn B9) when the vectors are real. Note that the inner product of two different vectors will in general be a complex number, but the length of a vector is always real and positive.

Appendix C. Operators

Operators are mathematical devices which take in a mathematical object of some kind and convert it to another (usually different) object of the same kind. In many elementary treatments of quantum mechanics for chemists, operators are depicted as acting on functions; for example, differentiation is an operator acting on a function to give another function.

Within the Dirac bra and ket formalism, operators act on kets to produce other kets:

$$\hat{A}|\psi\rangle = |\psi'\rangle. \tag{C1}$$

Operators are conventionally written with 'hats' on, as shown here, though the hats are sometimes dropped for brevity. In this text we only drop the hats in the case of NMR *product operators* where it is traditional not to use them. As before, matrices will be written in ***bold italic***, where $\boldsymbol{A}$ is the matrix corresponding to $\hat{A}$.

As these kets may be thought of as complex vectors, operators can be described using complex square matrices. In many cases it is simplest to discuss the properties of operators in terms of the properties of the corresponding matrices. More generally, the fact that quantum mechanical operators can be represented by matrices indicates that they must be *linear operators*. As was the case for matrices this allows operator calculations to be broken down into simpler pieces, and then reassembled at the end.

The product of two operators, $\hat{A}_1\hat{A}_2$, is equivalent to operating first with the operator $\hat{A}_2$ and then with the operator $\hat{A}_1$. More simply, the matrix corresponding to $\hat{A}_1\hat{A}_2$ is the matrix product $\boldsymbol{A}_1\boldsymbol{A}_2$. Just as the order in which matrices are multiplied is important (i.e. $\boldsymbol{A}_1\boldsymbol{A}_2 \neq \boldsymbol{A}_2\boldsymbol{A}_1$ in general), the effect of two operators will usually depend on the order in which they are applied. This effect is summarized by the *operator commutator*,

$$\left[\hat{A}_1,\hat{A}_2\right] = \hat{A}_1\hat{A}_2 - \hat{A}_2\hat{A}_1. \tag{C2}$$

If the operator commutator is zero, then the operators are said to *commute*, and it does not matter in which order they are applied. Clearly, every operator commutes with itself, and so commutes with every power of itself; for example,

$$\left[\hat{A},\hat{A}^2\right] = \left[\hat{A},\hat{A}\hat{A}\right] = 0. \tag{C3}$$

The *inverse* of an operator $\hat{A}$ is written as $\hat{A}^{-1}$, and is defined such that

$$\hat{A}\hat{A}^{-1} = \hat{A}^{-1}\hat{A} = \hat{1}, \tag{C4}$$

where $\hat{1}$ is the unit or identity operator, which leaves every ket unchanged; the corresponding matrix is the identity matrix, $\boldsymbol{1}$, for which all the elements are zero except those on the diagonal which are all equal to one. While it may be difficult to determine an operator inverse, the matrix inverse is easily found using numerical techniques. Another important concept, most easily defined in terms of matrices, is the operator *adjoint*: the adjoint $\hat{A}^{\dagger}$ of an operator $\hat{A}$ is represented by the matrix $\boldsymbol{A}^{\dagger}$, obtained by taking the complex conjugate of the matrix transpose of $\boldsymbol{A}$.

Note that the identity operator commutes with everything.

The adjoint and inverse can be used to define two important groups of matrices, and hence operators. A matrix is said to be *Hermitian*, or self-adjoint, if $\boldsymbol{A} = \boldsymbol{A}^{\dagger}$, while it is *unitary* if $\boldsymbol{A}^{-1} = \boldsymbol{A}^{\dagger}$. Operators which correspond to *observables* (such as the Hamiltonian, which corresponds to the total energy) are always Hermitian. A matrix is said to be *normal* if $\boldsymbol{A}\boldsymbol{A}^{\dagger} = \boldsymbol{A}^{\dagger}\boldsymbol{A}$, and all Hermitian and unitary matrices are obviously normal.

Another important property is the existence of *eigenvectors* of operators; these have the property that they are scaled by the operator, but otherwise left unchanged. Eigenvectors, and their corresponding *eigenvalues*, will be explored in more detail in Appendix D. A key result, the *spectral decomposition theorem*, states that every normal $n \times n$ matrix has a *complete set* of n eigenvalues and eigenvectors.

Appendix D. Matrix exponentials

When a matrix is diagonal (i.e. all the elements off the main diagonal are zero) it is particularly simple to evaluate matrix powers. Consider a general 2×2 diagonal matrix

$$\boldsymbol{D} = \begin{pmatrix} a & 0 \\ 0 & b \end{pmatrix} \tag{D1}$$

for which

$$\boldsymbol{D}^2 = \begin{pmatrix} a & 0 \\ 0 & b \end{pmatrix}\begin{pmatrix} a & 0 \\ 0 & b \end{pmatrix} = \begin{pmatrix} a^2 & 0 \\ 0 & b^2 \end{pmatrix}. \tag{D2}$$

Similar expressions can be written for any power of $\boldsymbol{D}$, and thus for the matrix exponential (which is a sum of powers of $\boldsymbol{D}$):

$$\mathrm{e}^{\boldsymbol{D}} = \boldsymbol{1} + \boldsymbol{D} + \frac{\boldsymbol{D}^2}{2!} + \frac{\boldsymbol{D}^3}{3!} + \cdots = \begin{pmatrix} \mathrm{e}^a & 0 \\ 0 & \mathrm{e}^b \end{pmatrix}. \tag{D3}$$

Most matrices, such as $\boldsymbol{I_x}$, are not diagonal, and this simple approach cannot be applied. However, it is possible to do something very similar by *diagonalizing* the matrix.

A matrix $\boldsymbol{M}$ can be diagonalized by finding matrices $\boldsymbol{S}$ and $\boldsymbol{\Lambda}$ such that

$$\boldsymbol{M} = \boldsymbol{S}\boldsymbol{\Lambda}\boldsymbol{S}^{-1} \tag{D4}$$

where $\boldsymbol{\Lambda}$ is diagonal. Then

$$\boldsymbol{M}^2 = \boldsymbol{S}\boldsymbol{\Lambda}\boldsymbol{S}^{-1}\boldsymbol{S}\boldsymbol{\Lambda}\boldsymbol{S}^{-1} = \boldsymbol{S}\boldsymbol{\Lambda}\boldsymbol{\Lambda}\boldsymbol{S}^{-1} = \boldsymbol{S}\boldsymbol{\Lambda}^2\boldsymbol{S}^{-1}, \tag{D5}$$

Using $\boldsymbol{S}^{-1}\boldsymbol{S} = \boldsymbol{1}$.

and so on. Thus,

$$\mathrm{e}^{\boldsymbol{M}} = \boldsymbol{S}\mathrm{e}^{\boldsymbol{\Lambda}}\boldsymbol{S}^{-1}. \tag{D6}$$

A matrix can be diagonalized by finding its eigenvectors and eigenvalues. Suppose we have an $n \times n$ square matrix $\boldsymbol{M}$ and a vector $\boldsymbol{x}$ such that

$$\boldsymbol{M}\boldsymbol{x} = \lambda\boldsymbol{x}, \tag{D7}$$

where λ is a number; $\boldsymbol{x}$ is said to be an eigenvector of $\boldsymbol{M}$ with eigenvalue λ. As long as $\boldsymbol{M}$ is normal there will be n eigenvectors, each with its own eigenvalue. Once the set of eigenvalues and eigenvectors has been found the matrices $\boldsymbol{S}$ and $\boldsymbol{\Lambda}$ can readily be determined: $\boldsymbol{\Lambda}$ contains the eigenvalues of $\boldsymbol{M}$ on its diagonal (and zeroes everywhere else), and the columns of $\boldsymbol{S}$ are the corresponding eigenvectors.

Recall that $\boldsymbol{M}$ is normal if $\boldsymbol{M}\boldsymbol{M}^\dagger = \boldsymbol{M}^\dagger\boldsymbol{M}$; see Appendix C.

One way to determine them is to rewrite the eigenvalue equation as

$$(\boldsymbol{M} - \lambda\boldsymbol{1})\boldsymbol{x} = \boldsymbol{0}, \tag{D8}$$

where $\boldsymbol{0}$ is a vector consisting entirely of zeroes and $\boldsymbol{1}$ is the unit matrix. This set of n simultaneous equations only has non-trivial solutions if the determinant of the coefficients disappears, i.e. if

The trivial solution is $\boldsymbol{x} = \boldsymbol{0}$. Note that $\boldsymbol{0}$ is sometimes just written as 0.

$$|\boldsymbol{M} - \lambda\boldsymbol{1}| = 0. \tag{D9}$$

This expression is called the *secular equation*; expansion of the determinant gives a polynomial with n roots, corresponding to the n eigenvalues. Given the list of eigenvalues, the corresponding eigenvectors can be found by solving Eqn D8. The resulting eigenvectors are only defined up to multiplication by an arbitrary scalar; methods to partially resolve this ambiguity are discussed further in Appendix E, which also shows an example calculation in detail.

Note that the polynomial defined by Eqn D9 can have *repeated roots*, so the n eigenvalues do not have to be unique. Two or more eigenvectors with the same eigenvalue are said to be *degenerate*, and this leads to a further ambiguity in the definition of the eigenvectors, as any weighted sum of two degenerate eigenvectors is also an eigenvector with the same eigenvalue. Once again there are standard methods to partially resolve these ambiguities, and a suitable set of eigenvectors can always be found.

This method works well for analytical calculations with small matrices, but for large matrices it swiftly becomes complicated. With moderate sized matrices it can be very helpful to use symbolic algebra packages such as *Mathematica* to do much of the brute-force work of calculations: such packages can be used to find eigenvalues and eigenvectors, or just to evaluate matrix exponentials directly. However beyond 4×4 matrices such packages may be unable to help much, as there is no general solution to quintic or higher secular equations.

Details of *Mathematica* and the other computer packages discussed here can be found at the end of the Bibliography.

For large matrices it is usually necessary to use numerical methods. These can be found in many texts on numerical analysis, and many packages are available, including commercial implementations such as *Matlab*, free implementations such as Gnu *Octave*, and packages extending other languages such as the *NumPy* and *SciPy* extensions for Python. Numerical methods can be used to find the eigenvalues and eigenvectors of a matrix, but matrix exponentials can also be calculated directly using more sophisticated versions of the series expansion shown in Eqn D3. Many packages have been specially developed for NMR simulation, notably the *Spinach* library for *Matlab* and the *SpinDynamica* library for *Mathematica*.

Another way of thinking about diagonalizing a matrix $\boldsymbol{M}$ is that it is equivalent to writing $\boldsymbol{M}$ in its own *eigenbasis*, i.e. a basis made up of orthonormal eigenvectors of $\boldsymbol{M}$; this explains why the matrix $\boldsymbol{S}$ is formed from the eigenvectors of $\boldsymbol{M}$, and why $\boldsymbol{\Lambda}$ contains the corresponding eigenvalues. Thinking about a matrix in its eigenbasis can immediately reveal important properties. For example, a Hermitian matrix is equal to its own adjoint, and this must also be true for the diagonal matrix $\boldsymbol{\Lambda}$; this immediately leads to the conclusion that the individual eigenvalues must equal their own conjugates,

$$\lambda = \lambda^*, \tag{D10}$$

and thus the eigenvalues of a Hermitian matrix must all be real. A similar argument shows that the eigenvalues of a unitary matrix must be complex numbers with absolute value 1, that is

$$\lambda\lambda^* = 1, \tag{D11}$$

and so

$$\lambda = e^{ia}, \tag{D12}$$

where a is a real number. From these two results it is easy to see that the matrix exponential of a Hermitian matrix must be a unitary matrix.

Appendix E. The matrix exponential of $\hat{I}_x$

The matrix exponential of $\hat{I}_x$ can be determined using the approach described in Appendix D. Let $\boldsymbol{M}$ be $k\boldsymbol{I}_x$, where k is some scalar number. The secular equation is

$$\begin{vmatrix} -\lambda & \frac{1}{2}k \\ \frac{1}{2}k & -\lambda \end{vmatrix} = \lambda^2 - \tfrac{1}{4}k^2 = 0. \tag{E1}$$

Thus the two eigenvalues are $\lambda = \pm k/2$.

The eigenvectors may be deduced by substituting each of the eigenvalues in turn into the eigenvalue equation, $(\boldsymbol{M} - \lambda\boldsymbol{1})\boldsymbol{x} = \boldsymbol{0}$. For $\lambda = k/2$,

$$\begin{pmatrix} -\frac{1}{2}k & \frac{1}{2}k \\ \frac{1}{2}k & -\frac{1}{2}k \end{pmatrix}\begin{pmatrix} x_1 \\ x_2 \end{pmatrix} = \begin{pmatrix} 0 \\ 0 \end{pmatrix}, \tag{E2}$$

which has the solution $x_1 = x_2$. Note that the eigenvector is not completely defined: we could multiply x_1 and x_2 by any number and still have an eigenvector. One common convention is to normalize the eigenvectors to unit length; for the case considered above, this gives $x_1 = x_2 = 1/\sqrt{2}$. The second eigenvector can be determined in the same way by substituting $\lambda = -k/2$, giving $x_1 = 1/\sqrt{2}, x_2 = -1/\sqrt{2}$. Note that even after normalization some ambiguity remains as we could multiply x_1 and x_2 by -1, or more generally by any complex number of the form $e^{i\phi}$, as such numbers have magnitude 1. In general there is no way to resolve this ambiguity, but fortunately this ambiguity has no practical consequences.

The matrices $\boldsymbol{\Lambda}$ and $\boldsymbol{S}$ are thus (Appendix D)

$$\boldsymbol{\Lambda} = \begin{pmatrix} \frac{1}{2}k & 0 \\ 0 & -\frac{1}{2}k \end{pmatrix}, \qquad \boldsymbol{S} = \frac{1}{\sqrt{2}}\begin{pmatrix} 1 & 1 \\ 1 & -1 \end{pmatrix}. \tag{E3}$$

The matrix $\boldsymbol{S}$ is unitary (this is a general property of matrices of normalized eigenvectors), so $\boldsymbol{S}^{-1} = \boldsymbol{S}^{\dagger}$, and as $\boldsymbol{S}$ happens to be real, $\boldsymbol{S}^{\dagger} = \boldsymbol{S}^{\mathrm{T}}$. Now we can calculate the matrix exponential of $\boldsymbol{M}$, with $k = \pm i\omega_1 t$ (as in Section 7.7):

$$e^{\pm i\omega_1 t \boldsymbol{I}_x} = \boldsymbol{S} e^{\boldsymbol{\Lambda}} \boldsymbol{S}^{-1} = \frac{1}{2}\begin{pmatrix} 1 & 1 \\ 1 & -1 \end{pmatrix}\begin{pmatrix} e^{\pm i\omega_1 t/2} & 0 \\ 0 & e^{\mp i\omega_1 t/2} \end{pmatrix}\begin{pmatrix} 1 & 1 \\ 1 & -1 \end{pmatrix} \tag{E4}$$

$$= \begin{pmatrix} \cos(\frac{1}{2}\omega_1 t) & \pm i \sin(\frac{1}{2}\omega_1 t) \\ \pm i \sin(\frac{1}{2}\omega_1 t) & \cos(\frac{1}{2}\omega_1 t) \end{pmatrix}.$$

N.B. $\boldsymbol{S}^{-1} = \boldsymbol{S}^{\dagger} = \boldsymbol{S}^{\mathrm{T}} = \boldsymbol{S}$ in this case; in general only the first of these will be true.

Choosing the minus sign reproduces Eqn 7.53.

A more elegant proof is possible here because $\boldsymbol{I}_x$ has the special property that its square is proportional to the unit matrix, $\boldsymbol{1}$ (Eqn 7.28). If a matrix $\boldsymbol{A}$ is such that $\boldsymbol{A}^2 = \boldsymbol{1}$ then

$$e^{ia\boldsymbol{A}} \equiv \boldsymbol{1} + ia\boldsymbol{A} + \frac{i^2 a^2 \boldsymbol{A}^2}{2!} + \frac{i^3 a^3 \boldsymbol{A}^3}{3!} + \cdots \tag{E5}$$

may be rewritten, using $\boldsymbol{A} = \boldsymbol{A}^3 = \boldsymbol{A}^5 = \cdots$ and $\boldsymbol{1} = \boldsymbol{A}^2 = \boldsymbol{A}^4 = \cdots$, as

$$\boldsymbol{1}\left(1 - \frac{a^2}{2!} + \frac{a^4}{4!} - \cdots\right) + i\boldsymbol{A}\left(a - \frac{a^3}{3!} + \frac{a^5}{5!} - \cdots\right). \tag{E6}$$

The two series in parentheses are simply the expansions of $\cos(a)$ and $\sin(a)$, so that

$$e^{ia\boldsymbol{A}} = \boldsymbol{1}\cos a + i\boldsymbol{A}\sin a. \tag{E7}$$

To evaluate $\exp[\pm i\omega_1 t \boldsymbol{I}_x]$, we replace $\boldsymbol{A}$ by $2\boldsymbol{I}_x$ (because $4\boldsymbol{I}_x^2 = \boldsymbol{1}$) and a by $\pm\omega_1 t/2$ to obtain

$$e^{\pm i\omega_1 t \boldsymbol{I}_x} = \boldsymbol{1}\cos\left(\tfrac{1}{2}\omega_1 t\right) \pm i2\boldsymbol{I}_x \sin\left(\tfrac{1}{2}\omega_1 t\right) = \begin{pmatrix} \cos(\frac{1}{2}\omega_1 t) & \pm i\sin(\frac{1}{2}\omega_1 t) \\ \pm i\sin(\frac{1}{2}\omega_1 t) & \cos(\frac{1}{2}\omega_1 t) \end{pmatrix}. \tag{E8}$$

Appendix F. The rotating frame

The rotating-frame transformation plays a central role in the theory of many NMR experiments, greatly simplifying calculations by removing the continual rotation at the Larmor frequency which clutters things up in the laboratory frame.

Radiofrequency irradiation of an NMR sample creates a linearly oscillating magnetic field, which can be taken to lie along the x-axis so that

$$\hat{H} = \omega_0 \hat{I}_z + 2\omega_1 \cos(\omega_{\text{rf}} t + \phi)\hat{I}_x \tag{F1}$$

where ϕ, the phase of the radiation, will define the direction of the radiation in the rotating frame. For simplicity we assume $\phi = 0$ from here on. The reason for including a factor of 2 in the field strength will soon become clear.

This cosine modulation can be decomposed into the sum of two counter-rotating terms

$$\cos\omega_{\text{rf}} t = \tfrac{1}{2}\left(e^{+\mathrm{i}\omega_{\text{rf}} t} + e^{-\mathrm{i}\omega_{\text{rf}} t}\right), \tag{F2}$$

and similarly the oscillating part of the Hamiltonian can be decomposed into two counter-rotating fields:

$$\hat{H} = \omega_0 \hat{I}_z + \omega_1\left(\cos\omega_{\text{rf}} t\,\hat{I}_x + \sin\omega_{\text{rf}} t\,\hat{I}_y\right) + \omega_1\left(\cos\omega_{\text{rf}} t\,\hat{I}_x - \sin\omega_{\text{rf}} t\,\hat{I}_y\right). \tag{F3}$$

If the oscillating field is nearly resonant with the NMR transition, one component rotates at a frequency ω_{rf}, close to the Larmor frequency, while the other is at $-\omega_{\text{rf}}$. The latter will have little effect and can be ignored. Thus, the effective Hamiltonian in the laboratory frame can be written as

$$\hat{H} = \omega_0 \hat{I}_z + \omega_1\left(\cos\omega_{\text{rf}} t\,\hat{I}_x + \sin\omega_{\text{rf}} t\,\hat{I}_y\right), \tag{F4}$$

where the choice $\phi = 0$ for the phase of the radiofrequency radiation ensures that the effective Hamiltonian initially lies along the x-axis of the laboratory frame. As this Hamiltonian varies rapidly with time it is difficult to calculate its effect on a spin system. However, as the time variation has a simple regular form it is possible to remove it by transferring to a coordinate system in synchrony with the radiofrequency field as outlined in Section 1.3. This may be seen as follows.

The evolution of a wave function under a Hamiltonian is given by the time-dependent Schrödinger equation,

$$\frac{\mathrm{d}}{\mathrm{d}t}|\psi\rangle = -\mathrm{i}\hat{H}|\psi\rangle. \tag{F5}$$

Let us define a modified wave function, $|\psi_{\text{R}}\rangle$, related to $|\psi\rangle$ by rotation around the z-axis at a rate ω_{rf},

$$|\psi_{\text{R}}\rangle = \mathrm{e}^{\mathrm{i}\omega_{\text{rf}} t \hat{I}_z}|\psi\rangle. \tag{F6}$$

The time-dependence of $|\psi_R\rangle$ can be written in terms of a modified Hamiltonian as follows:

$$\begin{aligned}\frac{d}{dt}|\psi_R\rangle &= i\omega_{rf}\hat{I}_z e^{i\omega_{rf}t\hat{I}_z}|\psi\rangle + e^{i\omega_{rf}t\hat{I}_z}\frac{d}{dt}|\psi\rangle \\ &= i\omega_{rf}\hat{I}_z|\psi_R\rangle + e^{i\omega_{rf}t\hat{I}_z}(-i\hat{H})|\psi\rangle \\ &= -i\left[-\omega_{rf}\hat{I}_z|\psi_R\rangle + e^{i\omega_{rf}t\hat{I}_z}\hat{H}e^{-i\omega_{rf}t\hat{I}_z}|\psi_R\rangle\right] \\ &= -i\left[-\omega_{rf}\hat{I}_z + e^{i\omega_{rf}t\hat{I}_z}\hat{H}e^{-i\omega_{rf}t\hat{I}_z}\right]|\psi_R\rangle \\ &\equiv -i\hat{H}_R|\psi_R\rangle. \end{aligned} \quad (F7)$$

The third line follows from the second using $|\psi\rangle = \exp[-i\omega_{rf}t\hat{I}_z]|\psi_R\rangle$.

The rotating frame Hamiltonian $\hat{H}_R$ is made up of two parts: the original Hamiltonian subjected to the same rotation as the wave function, and an additional term, $-\omega_{rf}\hat{I}_z$, lying along the rotation axis. For the laboratory frame Hamiltonian (Eqn F4),

$$\begin{aligned}\hat{H}_R &= -\omega_{rf}\hat{I}_z + e^{i\omega_{rf}t\hat{I}_z}\left[\omega_0\hat{I}_z + \omega_1\left(\cos\omega_{rf}t\,\hat{I}_x + \sin\omega_{rf}t\,\hat{I}_y\right)\right]e^{-i\omega_{rf}t\hat{I}_z} \\ &= -\omega_{rf}\hat{I}_z + e^{i\omega_{rf}t\hat{I}_z}\omega_0\hat{I}_z e^{-i\omega_{rf}t\hat{I}_z} + \omega_1 e^{i\omega_{rf}t\hat{I}_z}\left(\cos\omega_{rf}t\,\hat{I}_x + \sin\omega_{rf}t\,\hat{I}_y\right)e^{-i\omega_{rf}t\hat{I}_z} \\ &= -\omega_{rf}\hat{I}_z + \omega_0\hat{I}_z + \omega_1\hat{I}_x \\ &= (\omega_0 - \omega_{rf})\hat{I}_z + \omega_1\hat{I}_x, \end{aligned} \quad (F8)$$

where the second line follows by linearity, and the middle term of the third line arises because exponentials of operators commute with the underlying operator. The third term is most simply calculated using matrix forms, as

$$\begin{aligned} e^{i\theta \mathbf{I_z}}\left(\cos\theta\,\mathbf{I_x} + \sin\theta\,\mathbf{I_y}\right)e^{-i\theta \mathbf{I_z}} &= \begin{pmatrix} e^{i\theta/2} & 0 \\ 0 & e^{-i\theta/2}\end{pmatrix}\frac{1}{2}\begin{pmatrix} 0 & e^{-i\theta} \\ e^{i\theta} & 0\end{pmatrix}\begin{pmatrix} e^{-i\theta/2} & 0 \\ 0 & e^{i\theta/2}\end{pmatrix} \\ &= \begin{pmatrix} e^{i\theta/2} & 0 \\ 0 & e^{-i\theta/2}\end{pmatrix}\frac{1}{2}\begin{pmatrix} 0 & e^{-i\theta/2} \\ e^{i\theta/2} & 0\end{pmatrix} \\ &= \frac{1}{2}\begin{pmatrix} 0 & 1 \\ 1 & 0\end{pmatrix} \\ &= \mathbf{I_x}. \end{aligned} \quad (F9)$$

This result can also be deduced using the methods in Section 8.6.

Thus the time dependence of the radiofrequency field has been removed and the precession frequency around the $\mathbf{B}_0$ direction has been reduced from ω_0 to $\omega_0 - \omega_{rf} = \Omega$. If the radiofrequency field ω_1 is much larger than the resonance offset Ω, the rotating frame Hamiltonian becomes particularly simple:

$$\hat{H}_R \approx \omega_1\hat{I}_x. \quad (F10)$$

This result holds exactly, of course, if the radiofrequency field is in exact resonance with the Larmor frequency, $\omega_0 = \omega_{rf}$.

Appendix G. Operator descriptions of pure states

As discussed in Section 8.4, the thermal equilibrium state of an NMR spin system is a mixed state, rather than a pure state of the kind used in many elementary treatments of quantum mechanics. While pure states can occur in some exotic *hyperpolarized* systems, these are rarely seen in everyday NMR.

It is, however, sometimes useful to describe pure spin states, and so it is useful to see how to do this. Here we only consider pure states of single isolated spin-$\frac{1}{2}$ nuclei; pure states in systems with two or more spins are discussed in Appendix J.

The simplest states to describe are the eigenstates of the Hamiltonian, $|\alpha\rangle$ and $|\beta\rangle$. The density operator for the pure state $|\alpha\rangle$ is $|\alpha\rangle\langle\alpha|$, with the density matrix

$$\begin{aligned}\rho_{\alpha\alpha} &= \begin{pmatrix} 1 & 0 \\ 0 & 0 \end{pmatrix} \\ &= \begin{pmatrix} \frac{1}{2} & 0 \\ 0 & \frac{1}{2} \end{pmatrix} + \begin{pmatrix} \frac{1}{2} & 0 \\ 0 & -\frac{1}{2} \end{pmatrix} \\ &= \tfrac{1}{2}\mathbf{1} + \mathbf{I_z}.\end{aligned} \tag{G1}$$

Similarly, for $|\beta\rangle$

$$\begin{aligned}\rho_{\beta\beta} &= \begin{pmatrix} 0 & 0 \\ 0 & 1 \end{pmatrix} \\ &= \begin{pmatrix} \frac{1}{2} & 0 \\ 0 & \frac{1}{2} \end{pmatrix} - \begin{pmatrix} \frac{1}{2} & 0 \\ 0 & -\frac{1}{2} \end{pmatrix} \\ &= \tfrac{1}{2}\mathbf{1} - \mathbf{I_z}.\end{aligned} \tag{G2}$$

Note that these results can also be obtained from the expression for a thermal equilibrium state by setting the polarization, $p_\alpha - p_\beta$, equal to 1 for $|\alpha\rangle$ and -1 for $|\beta\rangle$, and the corresponding operator expressions,

$$\hat{I}_\alpha = |\alpha\rangle\langle\alpha| = \tfrac{1}{2}\hat{1} + \hat{I}_z, \quad \hat{I}_\beta = |\beta\rangle\langle\beta| = \tfrac{1}{2}\hat{1} - \hat{I}_z, \tag{G3}$$

See Ernst et al. (1987).

are usually called *polarization operators*. The corresponding expressions for product operators can then be obtained by dropping the 'hats' and replacing 1 with E:

$$I_\alpha = \tfrac{1}{2}E + I_z, \quad I_\beta = \tfrac{1}{2}E - I_z. \tag{G4}$$

It is in principle possible to write operator expressions for pure superposition states, but this is very rare in conventional NMR experiments.

Appendix H. Off-resonance and frequency-selective pulses

The propagator for an off-resonance pulse along the *x*-axis is given in Eqn 8.42, and here we consider its effects on a spin starting in the thermal equilibrium state $\hat{I}_z$. Repeating the calculations in Eqns 8.34 and 8.38 gives

$$\langle \hat{I}_x \rangle = \frac{f - f\cos\left(\sqrt{1+f^2}\,\omega_1 t\right)}{2\left(1+f^2\right)}$$
$$\langle \hat{I}_y \rangle = \frac{-\sin\left(\sqrt{1+f^2}\,\omega_1 t\right)}{2\sqrt{1+f^2}}$$
$$\langle \hat{I}_z \rangle = \frac{f^2 + \cos\left(\sqrt{1+f^2}\,\omega_1 t\right)}{2\left(1+f^2\right)} \tag{H1}$$

where the values depend on the *off-resonance fraction* $f = \Omega/\omega_1$. As expected, these reduce to the results in Eqn 8.38 when f goes to 0. To make further progress it is useful to write $\omega_1 t = \theta$, where θ is the nutation angle achieved by the pulse when it is applied on-resonance, and then to specialize to the cases $\theta = \pi/2$ (a 90°_x pulse) and $\theta = \pi$ (a 180°_x pulse).

The results for a 90°_x pulse are shown in Fig. H1. For small values of f, roughly $|f| \leq 1$, the pulse is effective at transferring the initial *z*-magnetization into the *xy* plane, but unless f is very small this magnetization is not aligned along the *y*-axis, but also contains an *x* component. This will result in a phase error in the corresponding NMR spectrum. For large values of f the magnetization is largely left along the *z*-axis, although a small component in the *xy* plane will normally be produced.

For small values of f, this phase error will be linear in the offset frequency, and so can be corrected using the normal phasing procedure described in Chapter 2.

The results for a 180°_x pulse, which are shown in Fig. H2, are broadly similar. For small values of f, the pulse acts as an effective inversion pulse, while for large values of f it has little effect. Between these extremes the pulse will transfer magnetization towards the *xy* plane, with a magnitude and phase that depends on f.

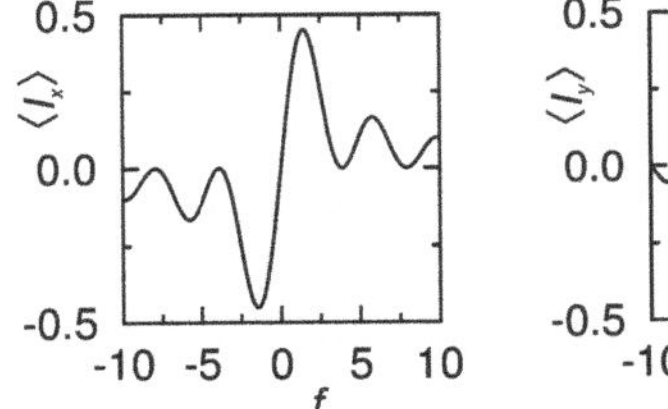

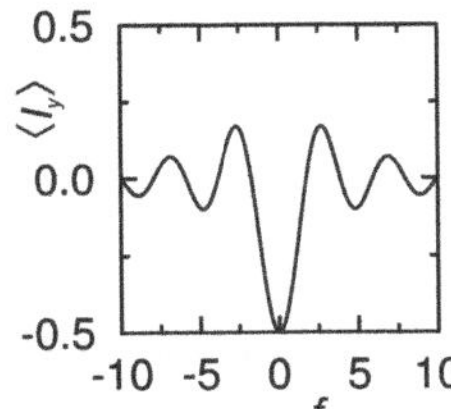

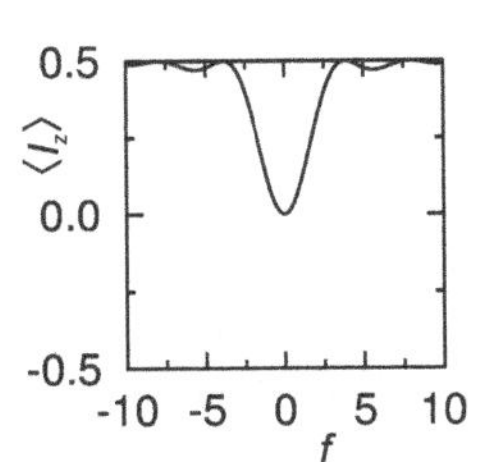

Fig. H1 The effects of an off-resonance 90°_x pulse on the thermal equilibrium state $\hat{I}_z$.

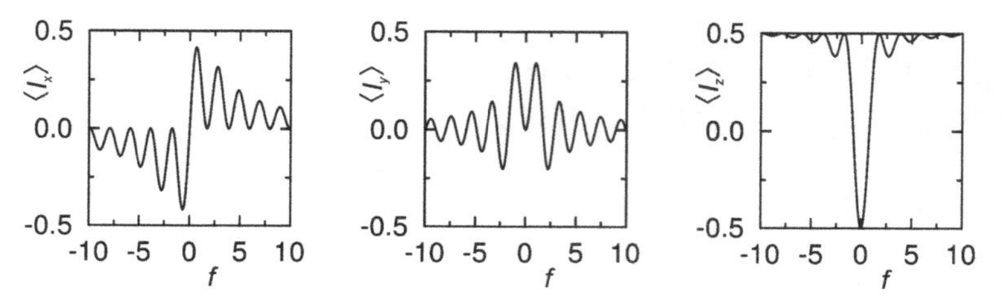

Fig. H2 The effects of an off-resonance 180°_x pulse on the thermal equilibrium state $\hat{I}_z$.

It is clear that a weak pulse can be used to preferentially excite one spin in a multi-spin system, but it is obvious from the figures that such a pulse is not really very selective. While off-resonance pulses have a large effect only when f is small, they continue to have a small effect even at very large values of f. This undesirable excitation of the off-resonance spins can be reduced by more sophisticated techniques, in particular *shaped pulses*. The excitation far off-resonance can ultimately be traced to the sharp edges of a pulse which is simply turned on and off, a so-called *square pulse*, and can be greatly reduced by smoothing these discontinuities, by smoothly varying the amplitude of the $\boldsymbol{B}_1$ field. More subtle variations in the amplitude and the phase of the field can be used to sculpt the excitation profile into any desired form, but the topic is too complex to discuss here.

For an introduction to shaped pulses, see Freeman *Spin Choreography* (1998).

Appendix I. Direct products

Direct products, also called tensor or Kronecker products, provide a convenient way of describing a system made up of two or more subsystems. If each of two subsystems is described by an $n \times n$ matrix, then the whole system requires an $n^2 \times n^2$ matrix to describe it completely. This matrix may be obtained as follows: a copy of the second matrix is multiplied by each element of the first matrix in turn, and the inner brackets are deleted:

$$\begin{pmatrix} a & b \\ c & d \end{pmatrix} \otimes \begin{pmatrix} \alpha & \beta \\ \gamma & \delta \end{pmatrix} \rightarrow \begin{pmatrix} a\begin{pmatrix} \alpha & \beta \\ \gamma & \delta \end{pmatrix} & b\begin{pmatrix} \alpha & \beta \\ \gamma & \delta \end{pmatrix} \\ c\begin{pmatrix} \alpha & \beta \\ \gamma & \delta \end{pmatrix} & d\begin{pmatrix} \alpha & \beta \\ \gamma & \delta \end{pmatrix} \end{pmatrix} = \begin{pmatrix} a\alpha & a\beta & b\alpha & b\beta \\ a\gamma & a\delta & b\gamma & b\delta \\ c\alpha & c\beta & d\alpha & d\beta \\ c\gamma & c\delta & d\gamma & d\delta \end{pmatrix}. \tag{I1}$$

Note that the order in which two matrices are multiplied together has no fundamental significance, but a consistent order must be used throughout a calculation; this corresponds to a consistent labelling of the two subsystems, I and S.

A complete list of the two-spin matrices is given in Table I1.

Table I1 Matrix representations of two-spin operators

$$I_x = \frac{1}{2}\begin{pmatrix} 0 & 0 & 1 & 0 \\ 0 & 0 & 0 & 1 \\ 1 & 0 & 0 & 0 \\ 0 & 1 & 0 & 0 \end{pmatrix} \quad I_y = \frac{1}{2}\begin{pmatrix} 0 & 0 & -i & 0 \\ 0 & 0 & 0 & -i \\ i & 0 & 0 & 0 \\ 0 & i & 0 & 0 \end{pmatrix} \quad I_z = \frac{1}{2}\begin{pmatrix} 1 & 0 & 0 & 0 \\ 0 & 1 & 0 & 0 \\ 0 & 0 & -1 & 0 \\ 0 & 0 & 0 & -1 \end{pmatrix}$$

$$S_x = \frac{1}{2}\begin{pmatrix} 0 & 1 & 0 & 0 \\ 1 & 0 & 0 & 0 \\ 0 & 0 & 0 & 1 \\ 0 & 0 & 1 & 0 \end{pmatrix} \quad S_y = \frac{1}{2}\begin{pmatrix} 0 & -i & 0 & 0 \\ i & 0 & 0 & 0 \\ 0 & 0 & 0 & -i \\ 0 & 0 & i & 0 \end{pmatrix} \quad S_z = \frac{1}{2}\begin{pmatrix} 1 & 0 & 0 & 0 \\ 0 & -1 & 0 & 0 \\ 0 & 0 & 1 & 0 \\ 0 & 0 & 0 & -1 \end{pmatrix}$$

$$2I_xS_x = \frac{1}{2}\begin{pmatrix} 0 & 0 & 0 & 1 \\ 0 & 0 & 1 & 0 \\ 0 & 1 & 0 & 0 \\ 1 & 0 & 0 & 0 \end{pmatrix} \quad 2I_xS_y = \frac{1}{2}\begin{pmatrix} 0 & 0 & 0 & -i \\ 0 & 0 & i & 0 \\ 0 & -i & 0 & 0 \\ i & 0 & 0 & 0 \end{pmatrix} \quad 2I_xS_z = \frac{1}{2}\begin{pmatrix} 0 & 0 & 1 & 0 \\ 0 & 0 & 0 & -1 \\ 1 & 0 & 0 & 0 \\ 0 & -1 & 0 & 0 \end{pmatrix}$$

$$2I_yS_x = \frac{1}{2}\begin{pmatrix} 0 & 0 & 0 & -i \\ 0 & 0 & -i & 0 \\ 0 & i & 0 & 0 \\ i & 0 & 0 & 0 \end{pmatrix} \quad 2I_yS_y = \frac{1}{2}\begin{pmatrix} 0 & 0 & 0 & -1 \\ 0 & 0 & 1 & 0 \\ 0 & 1 & 0 & 0 \\ -1 & 0 & 0 & 0 \end{pmatrix} \quad 2I_yS_z = \frac{1}{2}\begin{pmatrix} 0 & 0 & -i & 0 \\ 0 & 0 & 0 & i \\ i & 0 & 0 & 0 \\ 0 & -i & 0 & 0 \end{pmatrix}$$

$$2I_zS_x = \frac{1}{2}\begin{pmatrix} 0 & 1 & 0 & 0 \\ 1 & 0 & 0 & 0 \\ 0 & 0 & 0 & -1 \\ 0 & 0 & -1 & 0 \end{pmatrix} \quad 2I_zS_y = \frac{1}{2}\begin{pmatrix} 0 & -i & 0 & 0 \\ i & 0 & 0 & 0 \\ 0 & 0 & 0 & i \\ 0 & 0 & -i & 0 \end{pmatrix} \quad 2I_zS_z = \frac{1}{2}\begin{pmatrix} 1 & 0 & 0 & 0 \\ 0 & -1 & 0 & 0 \\ 0 & 0 & -1 & 0 \\ 0 & 0 & 0 & 1 \end{pmatrix}$$

$$ZQ_x = \frac{1}{2}\begin{pmatrix} 0 & 0 & 0 & 0 \\ 0 & 0 & 1 & 0 \\ 0 & 1 & 0 & 0 \\ 0 & 0 & 0 & 0 \end{pmatrix} \quad ZQ_y = \frac{1}{2}\begin{pmatrix} 0 & 0 & 0 & 0 \\ 0 & 0 & -i & 0 \\ 0 & i & 0 & 0 \\ 0 & 0 & 0 & 0 \end{pmatrix} \quad \frac{1}{2}E = \frac{1}{2}\begin{pmatrix} 1 & 0 & 0 & 0 \\ 0 & 1 & 0 & 0 \\ 0 & 0 & 1 & 0 \\ 0 & 0 & 0 & 1 \end{pmatrix}$$

$$DQ_x = \frac{1}{2}\begin{pmatrix} 0 & 0 & 0 & 1 \\ 0 & 0 & 0 & 0 \\ 0 & 0 & 0 & 0 \\ 1 & 0 & 0 & 0 \end{pmatrix} \quad DQ_y = \frac{1}{2}\begin{pmatrix} 0 & 0 & 0 & -i \\ 0 & 0 & 0 & 0 \\ 0 & 0 & 0 & 0 \\ i & 0 & 0 & 0 \end{pmatrix} \quad I \cdot S = \frac{1}{4}\begin{pmatrix} 1 & 0 & 0 & 0 \\ 0 & -1 & 2 & 0 \\ 0 & 2 & -1 & 0 \\ 0 & 0 & 0 & 1 \end{pmatrix}$$

Appendix J. Pure states of two-spin systems

As discussed in Appendix G, it is possible to describe pure states of single isolated spins as sums of operators

$$\hat{I}_\alpha = |\alpha\rangle\langle\alpha| = \tfrac{1}{2}\hat{1} + \hat{I}_z, \quad \hat{I}_\beta = |\beta\rangle\langle\beta| = \tfrac{1}{2}\hat{1} - \hat{I}_z. \tag{J1}$$

A similar approach can be used in systems with two or more spin-$\frac{1}{2}$ nuclei, but the descriptions become increasingly complicated. In a two-spin system the pure state $|\alpha\alpha\rangle$ has the density operator

$$|\alpha\alpha\rangle\langle\alpha\alpha| = \begin{pmatrix} 1 & 0 & 0 & 0 \\ 0 & 0 & 0 & 0 \\ 0 & 0 & 0 & 0 \\ 0 & 0 & 0 & 0 \end{pmatrix} = \tfrac{1}{2}\left(\tfrac{1}{2}\hat{1} + \hat{I}_z + \hat{S}_z + 2\hat{I}_z\hat{S}_z\right) \tag{J2}$$

which contains two-spin ordered population terms. (This expression can be confirmed by combining the density matrices for two-spin systems listed in Appendix I). The operators for other two-spin states can be obtained by changing the signs of the different components,

$$\begin{aligned} |\alpha\beta\rangle\langle\alpha\beta| &= \tfrac{1}{2}\left(\tfrac{1}{2}\hat{1} + \hat{I}_z - \hat{S}_z - 2\hat{I}_z\hat{S}_z\right), \\ |\beta\alpha\rangle\langle\beta\alpha| &= \tfrac{1}{2}\left(\tfrac{1}{2}\hat{1} - \hat{I}_z + \hat{S}_z - 2\hat{I}_z\hat{S}_z\right), \\ |\beta\beta\rangle\langle\beta\beta| &= \tfrac{1}{2}\left(\tfrac{1}{2}\hat{1} - \hat{I}_z - \hat{S}_z + 2\hat{I}_z\hat{S}_z\right). \end{aligned} \tag{J3}$$

Operators for three-spin states can be written in much the same way, and the number of components required doubles with every additional spin.

Appendix K. Some properties of commuting matrices

Here we prove a few useful properties of commuting matrices, i.e. matrices for which $[\boldsymbol{A},\boldsymbol{B}] = \boldsymbol{AB} - \boldsymbol{BA} = \boldsymbol{0}$ so that $\boldsymbol{AB} = \boldsymbol{BA}$. First:

$$\left[e^{\boldsymbol{A}},\boldsymbol{B}\right] = \left[e^{\boldsymbol{B}},\boldsymbol{A}\right] = \left[e^{\boldsymbol{A}},e^{\boldsymbol{B}}\right] = \boldsymbol{0} \tag{K1}$$

which follows from the fact that $e^{\boldsymbol{A}}$ is a sum of powers of $\boldsymbol{A}$, and $\boldsymbol{B}$ commutes with $\boldsymbol{A}$ and so with any power of $\boldsymbol{A}$. Since every matrix commutes with itself note that for *any* matrix $\boldsymbol{A}$

This property was used in Eqn 8.15.

$$\left[e^{\boldsymbol{A}},\boldsymbol{A}\right] = \boldsymbol{0}. \tag{K2}$$

Second:

$$e^{\boldsymbol{A}+\boldsymbol{B}} = e^{\boldsymbol{A}}e^{\boldsymbol{B}} = e^{\boldsymbol{B}}e^{\boldsymbol{A}}. \tag{K3}$$

This can be shown by expanding $e^{\boldsymbol{A}}$ and $e^{\boldsymbol{B}}$

$$\begin{aligned}
e^{\boldsymbol{A}}e^{\boldsymbol{B}} &= \left(\boldsymbol{1}+\boldsymbol{A}+\frac{\boldsymbol{A}^2}{2!}+\frac{\boldsymbol{A}^3}{3!}+\cdots\right)\left(\boldsymbol{1}+\boldsymbol{B}+\frac{\boldsymbol{B}^2}{2!}+\frac{\boldsymbol{B}^3}{3!}+\cdots\right)\\
&= \boldsymbol{1}+\boldsymbol{A}+\boldsymbol{B}+\frac{\boldsymbol{A}^2}{2!}+\boldsymbol{AB}+\frac{\boldsymbol{B}^2}{2!}+\frac{\boldsymbol{A}^3}{3!}+\frac{\boldsymbol{A}^2\boldsymbol{B}}{2!}+\frac{\boldsymbol{AB}^2}{2!}+\frac{\boldsymbol{B}^3}{3!}+\cdots\\
&= \boldsymbol{1}+(\boldsymbol{A}+\boldsymbol{B})+\frac{(\boldsymbol{A}+\boldsymbol{B})^2}{2!}+\frac{(\boldsymbol{A}+\boldsymbol{B})^3}{3!}+\cdots\\
&= e^{\boldsymbol{A}+\boldsymbol{B}},
\end{aligned} \tag{K4}$$

where the commutator $[\boldsymbol{A},\boldsymbol{B}]=\boldsymbol{0}$ has been used to write

$$\boldsymbol{A}^2+2\boldsymbol{AB}+\boldsymbol{B}^2=\boldsymbol{A}^2+\boldsymbol{AB}+\boldsymbol{BA}+\boldsymbol{B}^2\equiv(\boldsymbol{A}+\boldsymbol{B})^2, \tag{K5}$$

and so on.

Eqns K1–K3 lead to a considerable and very welcome simplification in some density matrix calculations. If the Hamiltonian is the sum of two commuting parts ($\boldsymbol{H}=\boldsymbol{H}_1+\boldsymbol{H}_2$), then the evolution predicted by the Liouville-von Neumann equation may be written

$$\begin{aligned}
\boldsymbol{\rho}(t) &= e^{-i\boldsymbol{H}t}\boldsymbol{\rho}(0)e^{i\boldsymbol{H}t}\\
&= e^{-i\boldsymbol{H}_1t}e^{-i\boldsymbol{H}_2t}\boldsymbol{\rho}(0)e^{i\boldsymbol{H}_2t}e^{i\boldsymbol{H}_1t}\\
&= e^{-i\boldsymbol{H}_2t}e^{-i\boldsymbol{H}_1t}\boldsymbol{\rho}(0)e^{i\boldsymbol{H}_1t}e^{i\boldsymbol{H}_2t},
\end{aligned} \tag{K6}$$

i.e. the evolution produced by $\boldsymbol{H}_1$ and $\boldsymbol{H}_2$ may be calculated sequentially, rather than simultaneously, and in any order, a property that holds however many commuting parts constitute the overall Hamiltonian. This result underlies the whole of the product operator formalism for weakly coupled spin systems.

Appendix L. Commutation relations in two-spin systems

As the behaviour of product operators can be deduced from their commutation relations, it is useful to have access to a table of commutators in two spin-systems. A complete table of such commutators is given in Table L1.

The commutators in Table L1 should be read as indicating that $\left[I_x, I_y\right]=iI_z$, and so on. The significance of the entries is that if a state corresponding to a particular column label evolves under a Hamiltonian corresponding to a row label, then the state evolves *towards* the corresponding table entry. If the table entry is zero then the operators commute and the state does not evolve under the given Hamiltonian.

These commutators can be calculated by brute force using the matrix representations in Table L1, but it is more sensible to use insight to simplify these calculations. Firstly, since $[\boldsymbol{A},\boldsymbol{B}] = -[\boldsymbol{B},\boldsymbol{A}]$, the table will be symmetric around its diagonal, which will be made up of zeroes. Secondly, since all operators involving only spin I commute with all operators involving only spin S, cross terms between these operators will all be zero.

Table L1 Operator commutators; the operator hats (ˆ) have been omitted here for simplicity.

	I_x	I_y	I_z	S_x	S_y	S_z	$2I_xS_x$	$2I_xS_y$	$2I_xS_z$	$2I_yS_x$	$2I_yS_y$	$2I_yS_z$	$2I_zS_x$	$2I_zS_y$	$2I_zS_z$
I_x	0	$+iI_z$	$-iI_y$	0	0	0	0	0	0	$+i2I_zS_x$	$+i2I_zS_y$	$+i2I_zS_z$	$-i2I_yS_x$	$-i2I_yS_y$	$-i2I_yS_z$
I_y	$-iI_z$	0	$+iI_x$	0	0	0	$-i2I_zS_x$	$-i2I_zS_y$	$-i2I_zS_z$	0	0	0	$+i2I_xS_x$	$+i2I_xS_y$	$+i2I_xS_z$
I_z	$+iI_y$	$-iI_x$	0	0	0	0	$+i2I_yS_x$	$+i2I_yS_y$	$+i2I_yS_z$	$-i2I_xS_x$	$-i2I_xS_y$	$-i2I_xS_z$	0	0	0
S_x	0	0	0	0	$+iS_z$	$-iS_y$	0	$+i2I_xS_z$	$-2iI_xS_y$	0	$+i2I_yS_z$	$-i2I_yS_y$	0	$+i2I_zS_z$	$-i2I_zS_y$
S_y	0	0	0	$-iS_z$	0	$+iS_x$	$-i2I_xS_z$	0	$+i2I_xS_x$	$-i2I_yS_z$	0	$+i2I_yS_x$	$-i2I_zS_z$	0	$+i2I_zS_x$
S_z	0	0	0	$+iS_y$	$-iS_x$	0	$+i2I_xS_y$	$-i2I_xS_x$	0	$+i2I_yS_y$	$-i2I_yS_x$	0	$+i2I_zS_y$	$-i2I_zS_x$	0
$2I_xS_x$	0	$+i2I_zS_x$	$-i2I_yS_x$	0	$+i2I_xS_z$	$-i2I_xS_y$	0	$+iS_z$	$-iS_y$	$+iI_z$	0	0	$-iI_y$	0	0
$2I_xS_y$	0	$+i2I_zS_y$	$-i2I_yS_y$	$-i2I_xS_z$	0	$+i2I_xS_x$	$-iS_z$	0	$+iS_x$	0	$+iI_z$	0	0	$-iI_y$	0
$2I_xS_z$	0	$+i2I_zS_z$	$-i2I_yS_z$	$+i2I_xS_y$	$-i2I_xS_x$	0	$+iS_y$	$-iS_x$	0	0	0	$+iI_z$	0	0	$-iI_y$
$2I_yS_x$	$-i2I_zS_x$	0	$+i2I_xS_x$	0	$+i2I_yS_z$	$-i2I_yS_y$	$-iI_z$	0	0	0	$+iS_z$	$-iS_y$	$+iI_x$	0	0
$2I_yS_y$	$-i2I_zS_y$	0	$+i2I_xS_y$	$-i2I_yS_z$	0	$+i2I_yS_x$	0	$-iI_z$	0	$-iS_z$	0	$+iS_x$	0	$+iI_x$	0
$2I_yS_z$	$-i2I_zS_z$	0	$+i2I_xS_z$	$+i2I_yS_y$	$-i2I_yS_x$	0	0	0	$-iI_z$	$+iS_y$	$-iS_x$	0	0	0	$+iI_x$
$2I_zS_x$	$+i2I_yS_x$	$-i2I_xS_x$	0	0	$+i2I_zS_z$	$-i2I_zS_y$	$+iI_y$	0	0	$-iI_x$	0	0	0	$+iS_z$	$-iS_y$
$2I_zS_y$	$+i2I_yS_y$	$-i2I_xS_y$	0	$-i2I_zS_z$	0	$+i2I_zS_x$	0	$+iI_y$	0	0	$-iI_x$	0	$-iS_z$	0	$+iS_x$
$2I_zS_z$	$+i2I_yS_z$	$-i2I_xS_z$	0	$+i2I_zS_y$	$-i2I_zS_x$	0	0	0	$+iI_y$	0	0	$-iI_x$	$+iS_y$	$-iS_x$	0

Commutators involving only a single two-spin operator are only slightly more complicated: note that

$$\begin{aligned}&\left[I_\alpha, 2I_\beta S_\gamma\right] = I_\alpha 2I_\beta S_\gamma - 2I_\beta S_\gamma I_\alpha\\ &= 2I_\alpha I_\beta S_\gamma - 2I_\beta I_\alpha S_\gamma\\ &= 2\left[I_\alpha, I_\beta\right] S_\gamma\end{aligned} \tag{L1}$$

The second line uses the facts that I spin operators commute with S spin operators and numbers commute with everything.

and so these results can be deduced from the corresponding single-spin commutators. However for commutators involving a pair of two-spin operators it is necessary to proceed with care. For example,

$$\begin{aligned}\left[2I_x S_x, 2I_x S_z\right] &= 4\left(I_x S_x I_x S_z - I_x S_z I_x S_x\right)\\ &= 4I_x^2\left[S_x, S_z\right]\\ &= -\mathrm{i}S_y\end{aligned} \tag{L2}$$

In a two-spin system $I_x^2 = 1/4$, and so on.

while

$$\begin{aligned}\left[2I_x S_y, 2I_y S_z\right] &= 4\left(I_x S_y I_y S_z - I_y S_z I_x S_y\right)\\ &= \left(2I_x I_y\right)\left(2S_y S_z\right) - \left(2I_y I_x\right)\left(2S_z S_y\right)\\ &= \left(\mathrm{i}I_z\right)\left(\mathrm{i}S_x\right) - \left(-\mathrm{i}I_z\right)\left(-\mathrm{i}S_x\right)\\ &= 0\end{aligned} \tag{L3}$$

Direct calculation shows that $I_x I_y = \frac{1}{2}\left[I_x, I_y\right]$, and so on. See Eqns 7.29 and 7.30.

Bibliography

Atkins, P. W. and Friedman, R. S. 2010. *Molecular Quantum Mechanics*, 5th ed. Oxford: Oxford University Press.

Cavanagh, J., Fairbrother, W. J., Palmer, A. G., III, and Skelton, N. J. 2006. *Protein NMR Spectroscopy*, 2nd ed. San Diego: Academic Press.

Ernst, R. R., Bodenhausen, G. and Wokaun, A. 1987. *Principles of Nuclear Magnetic Resonance in One and Two Dimensions*. Oxford: Oxford University Press.

Freeman, R. 1998. *Spin Choreography*. Oxford: Oxford University Press.

Hoch, J. C. and Stern, A. S. 1996. *NMR Data Processing*. New York: Wiley.

Hore, P. J. 2015. *Nuclear Magnetic Resonance*, 2nd ed. Oxford: Oxford University Press.

Jones, J. A. and Jaksch, D. J. 2012. *Quantum Information, Computation and Communication*. Cambridge: Cambridge University Press.

Kay, L. E., Ikura, M., Tschudin, R. and Bax, A. 1990. Three-dimensional triple resonance NMR spectroscopy of isotopically enriched proteins. *Journal of Magnetic Resonance* 89: 496.

Levitt, M. H. 1997. The signs of frequencies and phases in NMR. *Journal of Magnetic Resonance* 126: 164.

Levitt, M. H. 2008. *Spin Dynamics*, 2nd ed. New York: Wiley.

Shaka, A. J. and Keeler, J. 1987. Broadband spin decoupling in isotropic liquids. *Progress in NMR Spectroscopy* 19: 47.

Sørensen, O. W., Eich, G. W., Levitt, M. H., Bodenhausen, G. and Ernst, R. R. 1983. Product operator formalism for the description of NMR pulse experiments. *Progress in NMR Spectroscopy* 16: 163.

Susskind, L. and Friedman, A. 2014. *Quantum Mechanics: The Theoretical Minimum*. London: Allen Lane.

Computer Packages

GNU Octave: https://www.gnu.org/software/octave/

Mathematica: http://www.wolfram.com/

Matlab: http://www.mathworks.com/

SciPy: http://www.scipy.org/

Spinach: http://spindynamics.org/Spinach.php

SpinDynamica: http://www.spindynamica.soton.ac.uk/about/

Table of experiments

Abbreviation	Full name	Section
COSY	COrrelation SpectroscopY	5.2
DEPT	Distortionless Enhancement by Polarization Transfer	4.5
DQF-COSY	Double Quantum Filtered COSY	5.3
HMQC	Heteronuclear Multiple Quantum Correlation	5.6
HNCA	Hydrogen Nitrogen Carbon α correlation	5.8
HN(CO)CA	Hydrogen Nitrogen Carbon α correlation with indirect transfer via Carbonyl carbon	5.8
HSQC	Heteronuclear Single Quantum Correlation	5.7
HSQC-NOESY	Heteronuclear Single Quantum Correlation Nuclear Overhauser Effect SpectroscopY	5.8
HSQC-TOCSY	Heteronuclear Single Quantum Correlation TOtal Correlation SpectroscopY	5.8
INADEQUATE	Incredible Natural Abundance DoublE QUAntum Transfer Experiment	4.3
INEPT	Insensitive Nuclei Enhanced by Polarization Transfer	4.2
NOESY	Nuclear Overhauser Effect SpectroscopY	5.4
NOESY-HSQC	Nuclear Overhauser Effect SpectroscopY with Heteronuclear Single Quantum Correlation	5.8
ROESY	Rotating frame Overhauser Effect SpectroscopY	5.5
TOCSY	TOtal Correlation SpectroscopY	5.5
TOCSY-HSQC	TOtal Correlation SpectroscopY with Heteronuclear Single Quantum Correlation	5.8

Index